Climate Change
A Very Peculiar History®

With added urgency

'We know what is happening and what needs
to be done. Only you know if we did it.'

Message on a 2019 plaque by the former Ok glacier
in Iceland, the first lost to climate change

For my grandchildren and theirs –
that they have a world to enjoy.
DA

Editor: Nick Pierce
Artist: David Lyttleton

Published in Great Britain in MMXX by
Book House, an imprint of
The Salariya Book Company Ltd
25 Marlborough Place, Brighton BN1 1UB
www.salariya.com

ISBN: 978-1-912904-95-2

© The Salariya Book Company Ltd MMXX
All rights reserved. No part of this publication may be reproduced, stored in or introduced into a retrieval system or transmitted in any form, or by any means (electronic, mechanical, photocopying, recording or otherwise) without the written permission of the publisher. Any person who does any unauthorised act in relation to this publication may be liable to criminal prosecution and civil claims for damages.

A Very Peculiar History and all related titles and logos are registered trademarks of The Salariya Book Company.

1 3 5 7 9 8 6 4 2

A CIP catalogue record for this book is available
from the British Library.

Printed and bound in China.
Printed on paper from sustainable sources.

This book is sold subject to the conditions that it shall not, by way of trade or otherwise, be lent, resold, hired out, or otherwise circulated without the publisher's prior consent in any form or binding or cover other than that in which it is published and without similar condition being imposed on the subsequent purchaser.

Visit
www.salariya.com
for our online catalogue and
free fun stuff.

Climate Change
A Very Peculiar History®

With added urgency

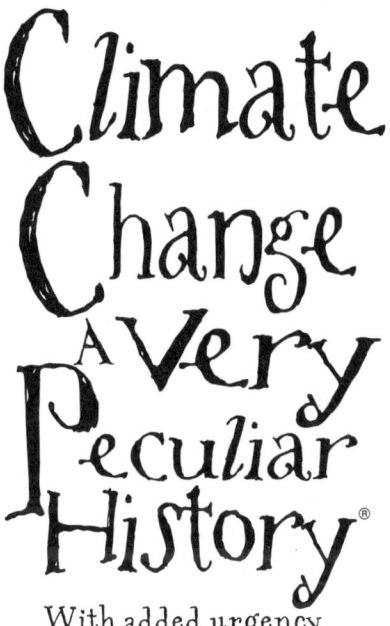

Written by
David Arscott

Created and designed by
David Salariya

BOOK HOUSE
a SALARIYA imprint

'Today we dumped another 70 million tons of global-warming pollution into the thin shell of atmosphere surrounding our planet, as if it were an open sewer. And tomorrow we will dump a slightly larger amount.'
Al Gore, former US vice president and environmental campaigner

'The climate change that is taking place because of increases in CO_2 concentration is largely irreversible for 1,000 years after emissions stop.'
Report by the National Oceanic and Atmospheric Administration

'The concept of global warming was invented by and for the Chinese in order to make US manufacturing non-competitive.'
Donald Trump, US president

'Our lives are in your hands.'
Greta Thunberg, young climate change activist

Contents

Introduction: Engines of destruction   7

1 Living in a greenhouse   25
2 What on Earth have we done?   39
3 Surviving the century   65
4 Speaking out   93
5 Desperate measures   125
6 The Doomsday Decade   167

Glossary   176
Climate Change Timeline   179
Index   184

' rapid, far-reaching and unprecedented changes in all aspects of society '

INTRODUCTION

ENGINES Of DESTRUCTION

On a drizzly September day in 1830 large crowds turned out for the opening of the world's first intercity railway line, carrying passengers between Liverpool and Manchester. The celebrations weren't entirely joyful – Lord Wellington, the unpopular prime minister, was pelted with rotten vegetables while, more seriously, the local MP William Huskisson carelessly stepped into the path of Robert Stephenson's prize-winning Rocket engine and died in agony hours later – but what the onlookers had witnessed was no less than a tipping point in the advance of the industrial revolution.

Who could possibly have foreseen the dire consequences of this engineering ingenuity? Certainly there were some who abominated the noisy, fiery monsters – and their plumes of acrid smoke and the specks of smut that lodged in an unwary eye were undeniably polluting. But a threat to life on Earth?

So it was to prove. Within a generation every major town in England was connected by the iron rails, and as the Victorians exported their technology around the world, so from continent to continent the burning of coal dispersed thick clouds of carbon into the atmosphere.

And this, of course, was only the beginning. After the age of steam came not only the invention of the petrol engine, but the harnessing of electricity, created from coal and oil and methane – the so-called fossil fuels – to power myriad inventions across the globe.

Scientists measuring the effects of this ruthless ransacking of the Earth's resources take for their base reading an ill-defined period in the 18th century before human activity disturbed the natural balance.

They measure the amount of carbon dioxide (CO_2) in the atmosphere as 'parts per million', or ppm, and here's what they've found: the concentration back in those tranquil days before the age of steam was around 280 ppm, whereas by 2019 it had risen to above 410 ppm.

The toxic build-up of these 'greenhouse gases' is relentless. The average annual increase has climbed from less than 1 ppm a year in the 1950s, when measurements began, to well over 2 ppm a year now.

Mauna Loa

To test air quality you need to be well away from direct sources of pollution. The observatory at Mauna Loa on Hawaii's Big Island – the world's chief benchmark site for carbon gas measurement – generally fits the bill as it's far away from its nearest continental landmass, although occasional volcanic activity causes contamination which has to be removed from the background data.

The graph of ppm readings is sometimes called the Keeling Curve after the observatory's first director, the American Charles Keeling.

Boffins united

In 1988 the United Nations Environmental Program and the World Meteorological Organization together set up the Intergovernmental Panel on Climate Change (IPCC), with the brief to issue full-scale reports every six years by scientists in the field from 195 member countries around the world, as well as a flow of separate reports on specific topics.

None of them makes comfortable reading, sometimes confirming earlier grim forecasts and sometimes finding that they had actually understated the potential dangers.

The IPCC issued a special report on global warming in October 2018, with more than 6,000 scientific references. It found that limiting the rise to 1.5°C (2.7°F) was possible but would require 'deep emissions reductions' and 'rapid, far-reaching and unprecedented changes in all aspects of society'.

One caveat: the IPCC reports pass through many hands before publication, not all of them scientific, the final document having to address political sensitivities in the countries involved.

A question of degree

All this pumping of carbon into the atmosphere – half a trillion tonnes of it since the start of the industrial revolution, with as much again absorbed by the oceans – has heated the Earth to troubling levels.

In 2015 close on two hundred nations* signed an agreement in Paris under the auspices of the United Nations aimed at restricting the increase in the global average temperature to 1.5°C (2.7°F) above pre-industrial levels.

This may not sound much, but we're already above the 1°C (1.8°F) mark, and once the figure rises to 2°C (3.6°F) we're warned to expect severe consequences in the way of (among much else) heatwaves, extreme storms, falls in crop production, a rise in sea levels and water shortages.

President Trump later decided to withdraw the US from the agreement, whereupon more than 3,600 leaders from America's states, cities, tribes, businesses and colleges formed the We Are Still In movement, vowing to support action in the spirit of Paris.

None of the Paris guidelines is legally binding, but they stimulated immediate measures to reduce carbon emissions:

- **Norway, the Netherlands and France decided to ban all petrol and diesel vehicles by 2025, 2030 and 2040 respectively.**

- **France ruled out using coal to create electricity after 2022.**

Slowly does it

In 2008 the UK had been the first country in the world to introduce a Climate Change Act, with a commitment to cut greenhouse gases by at least 80 per cent by 2050 compared with 1990.

The Conservative government's response to the Paris agreement was to refuse fresh UK action as 'existing targets are already stretching and the priority is to take action to meet them'. In 2019, however, a report by the independent Committee on Climate Change recommended a cutting of emissions to 'net-zero' by 2050 – and some campaign groups claimed that this still wasn't fast enough.

- The Netherlands parliament passed a law cutting greenhouse gases by 95 per cent by 2050.

- China set about increasing its natural forest areas from 9.07 to 13.67 billion cubic metres in 20 years, and promised a large scale conversion of farmland to forest. By 2017 it had increased its hydro, wind and solar power generation almost threefold.

- India agreed to reduce the 'emissions intensity' of its GDP by up to 35 per cent by 2030.

Doubters rightly questioned whether this was too little too late, but some kind of turning point had been reached.

Blowing hot and cold

We've all heard about ice ages, but periods of extreme heat are less well known. Scientists probing the distant past through ice cores, marine and lake sediments, cave deposits and tree rings have charted successive waves of hot and cold climate, and they've established that in the last million years glacial and interglacial cycles have occurred every 100,000 years.

During the most severe freeze-ups (and there have been at least five of them over millions of years) ice sheets 3km (1.9 miles) thick formed over large tracts of both the northern and southern hemispheres. In the most severe conditions, some 720 to 630 million years ago, they may even have reached the equator.

We're in a warmer, interglacial interlude of the Quaternary period right now, but our recent forebears had a relatively chilly time of it between the 14th and 19th centuries – a period that's been dubbed the Little Ice Age:

- Glaciers advanced in the Swiss Alps.

- The Baltic Sea, Manhattan Harbour and canals in the Netherlands froze over.

- In 1658 the Swedish army marched across the frozen sea to invade Denmark.

- 'Frost fairs' were held on the icebound River Thames in London. During the 1683–4 event the ice was measured at 28cm (11in) thick, while a highlight of the last one, in 1814, was the sight of an elephant taken for a walk across the frozen river at Blackfriars Bridge.

The tilting Earth

It's time to meet the Serbian genius Milutin Milanković (1879–1958), a mathematician, astronomer, civil engineer and climatologist, who first came up with an explanation for regular movements in global temperatures over long periods of time.

Those ice ages, he showed, were in part a result of the Earth's somewhat wayward journeys around the Sun – uneven in their progress, but nevertheless following distinct cycles which now bear his name.

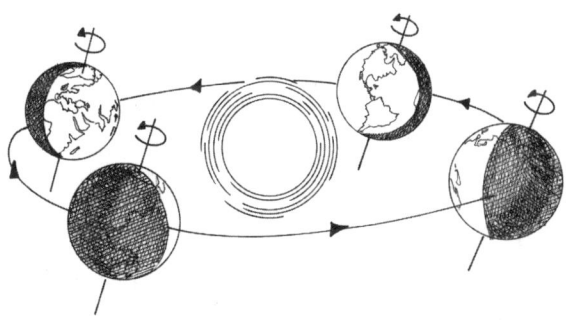

The Earth's axial tilt, spin and orbit around the Sun all affect the climate.

Here are the Milanković cycles:

1. The rotation cycle. The Earth spins like a toy spinning top, but it wobbles slightly on its axis. The wobbling is called 'precession'. As with the other two Milanković cycles, the wobble changes the amount of warmth received from the Sun by the Earth, with a cycle time of 26,000 years.

2. The tilt cycle. The Earth tilts, giving us our seasons, but the tilt changes. A greater tilt makes summers warmer and winters colder. A smaller tilt lessens the extremes of temperature. This cycle lasts 40,000 years.

3. The orbital cycle. The Earth's orbit around the Sun expands and contracts. Sometimes the orbit is more circular and sometimes it's more flattened and elliptical. The elliptical or circular shape of the orbit is called its 'eccentricity'. This cycle lasts 100,000 years.

It's a disturbing thought that, according to these calculations, we ought to be experiencing a cooler climate than we are. Instead it's getting warmer and warmer – *and if you've been paying attention you should guess why!*

The Earth's closeness to the Sun isn't the only cause of global warming. Another factor in the distant past was the build-up of CO_2 in the atmosphere because of massively erupting volcanoes, and today (as we've seen) we're pumping huge volumes of the stuff into the air every minute of every day. Volcanoes still play their part, but they can't compete with us.

Weather v climate

Weather is the rather unpredictable stuff we experience day by day, whereas climate is the underlying trend: it's sometimes called 'average weather'.

A snowstorm out of season is a freak but it's no argument against the fact that our world is warming. In fact the higher concentration of water vapour in the air contributes to ever fiercer storms of various kinds.

Do by all means blame the vagaries of air pressure, wind direction and so on for today's sudden downturn in the weather in your backyard, but *don't* confuse these temporary atmospheric conditions with the overall configurations of the climate.

And here's a frightening statistic: a few million years ago, when atmospheric concentrations of CO_2 in the atmosphere were similar to today's, the global temperature was between 2°C (3.6°F) and 3°C (5.4°F) higher than at the beginning of the industrial revolution, and the seas lapped the land all of 15–25m (50–80ft) above modern levels.

A warm welcome

Back in 1896 the Swedish physicist Svante Arrhenius (1859-1927) became the first scientist to calculate the effects of CO_2 on global warming – using his elaborate, exhausting workings-out by hand as a kind of therapy after a painful separation from his first wife.

Arrhenius concluded correctly that the ice ages had been triggered by a reduction of carbon in the atmosphere and that human activity was having the opposite effect in his own day. What he didn't realise was how greedy for fossil fuels later generations would become, and he therefore welcomed the Earth's increased toastiness as bringing a more benign climate to his native Scandinavia.

The Great Dying

The sobering fact is that the conditions sustaining life on Earth work in a precarious balance with one another. Mass extinctions have occurred five times in the past, long before humans walked the planet:

- **Ordovician-Silurian** (445 million years ago): 86 per cent of all species lost.

- **Late Devonian** (375 million years ago): 75 per cent of species lost.

- **End Permian** (251 million years ago): 96 per cent of species lost.

- **End Triassic** (200 million years ago): 80 per cent of species lost.

- **End Cretaceous** (66 million years ago): 76 per cent of species lost, including the dinosaurs.

These events had various causes (the crashing of an asteroid on the Earth is reckoned to have been a decisive factor in the last one), but it's the circumstances of the third extinction – known as 'The Great Dying' – that ought to touch a nerve with us.

It began with violent volcanic eruptions (stretching over all of two million years) from the Siberian Traps in Russia, propelling vast quantities of CO_2 into the atmosphere. As temperatures soared, the oceans acidified and stagnated. Anaerobic bacteria thrived on the lack of oxygen, emitting poisonous hydrogen sulfide, giving off (at least we weren't there to suffer) the horrible stench of rotten eggs.

It's estimated that more than half of all biological families and 80 per cent of all genera died out, and the regeneration of land-dwelling life took ten million years or more.

A sixth mass extinction?

Nothing quite as cataclysmic is forecast for our own immediate future, but with the globe heating up year by year, and with a wide variety of plant, animal, insect and reptile species already gone or under threat, we're living through a time of wildlife depletion sometimes referred to as the Anthropocene extinction.

Yes, that means we're causing it.

The lost summer

In 1815 Mount Tambora in what is now Indonesia spewed 100km^3 (24 cu miles) of toxic sulfates into the atmosphere – the most dramatic volcanic event in more than 1,600 years. Temperatures plummeted worldwide as vast swathes of ash veiled the sunlight, making 1816 'The Year Without a Summer'.

- The harvests failed in Britain and Ireland, and there were food riots in Europe.

- A persistent 'dry fog' enveloped parts of the eastern United States, crops withered and snow fell on frozen ground in June.

- In China the freeze killed trees and water buffalo, while floods destroyed rice crops.

- Torrential rains in India aggravated the spread of cholera from Bengal to as far away as Moscow.

Lord Byron caught the weirdness of it in his poem Darkness, written in 1816: 'Morn came and went – and came and brought no day,/And men forgot their passions in the dread of this their desolation.'

Black gold

Oil seeping through the ground has long been exploited for lighting and heating, and the 13th century traveller Marco Polo witnessed it gushing from a spring in Azerbaijan 'in such abundance that a hundred ships may load there at once', but the first oil well in the modern sense was sunk in Poland in 1854.

Its creator was the pharmacist Ignacy Łukasiewicz (1822–1882), a former political activist jailed for promoting Polish independence who later became both a wealthy businessman and a philanthropist.

In 1853, having discovered how to distil kerosene from seep oil, Łukasiewicz invented the modern kerosene lamp. The following year he opened the first of his many 'mines' in Bóbrka, near Gorlice – where a museum today displays two of his original hand-sunk oil wells, fondly named Frankek and Janina.

The first American oil well was the work of Edwin Drake at Titusville, Pennsylvania, in 1859. Together with a local blacksmith, William 'Uncle Billy' Smith, he built a steam-powered drilling machine and struck oil at a depth of 20m (66ft).

ENGINES OF DESTRUCTION

The overwhelming scientific evidence that our present climate change is both man-made (or 'anthropogenic') and dangerously intensifying is denied only by charlatans with an axe to grind or a profit to be made from heedlessly exploiting the Earth's natural resources.

We know what has to be done – and the price of inaction will be nothing short of disaster...

> Our green and pleasant troposphere is a place of delicate physical balances which we disturb at our peril

CHAPTER ONE

LIVING IN A GREENHOUSE

Life on Earth is something of a miracle. On our close companion, the Moon, temperatures sizzle at 113°C (253°F) in the glare of the Sun and plunge to an icy minus 158°C (minus 243°F) in its shadow. Its many craters testify to brutal bombardments by comets and asteroids.

What protects us from these horrors is a layer of gases so far not found to exist anywhere else in the universe. Our nurturing atmosphere includes the oxygen vital for life as we know it and a range of so-called greenhouse gases (or GHGs) which, absorbing and emitting radiant energy, wrap us in a warm comfort blanket.

Held fast to the Earth through gravity, the atmosphere is formed of several layers, the air increasingly rarified the higher you go.

- The TROPOSPHERE is where we live and where weather happens. It extends from the Earth's surface to about 10km (6.2 miles, or 33,000ft).

- Next up is the dry STRATOSPHERE, warmer at the top than the bottom, because it contains the ozone layer which absorbs the Sun's ultra violet radiation. Commercial airlines cruise in its lower levels, taking advantage of the reduced air density to save fuel.

- The MESOSPHERE is where most incoming meteors burn up. It extends from around 50 to 80km (30–50 miles)* – too high for aircraft and too low for orbital spacecraft.

- The International Space Station orbits in the chilly THERMOSPHERE, between 350 and 420km (220 and 260 miles) above the surface.

---

* *The bandwidths aren't precise, because they change with latitude and the seasons: higher in winter and at the tropics; lower in summer and at the poles.*

LIVING IN A GREENHOUSE

- **The 'final frontier' is the EXOSPHERE, the air so thin that it's difficult to define its boundary with outer space.**

As the diagram overleaf shows, our generally hospitable climate depends upon the interplay between the Sun's energy and the atmosphere's ability to both absorb and reflect it.

Up in a balloon

In 1902 two pioneers in 'aerology', the Frenchman Léon Teisserenc de Bort (1855-1913) and the German Richard Assmann (1845-1918), independently discovered the properties of the stratosphere by sending up measuring equipment to high altitudes in unmanned balloons.

Assmann, who invented a psychrometer to measure atmospheric humidity and temperatures, published a popular monthly magazine, *Das Wetter* (*The Weather*), and was awarded the Buys Ballot Medal by the Royal Netherlands Academy of Sciences. Teisserenc is remembered in the names of two craters, one on the Moon and the other on Mars.

The Balance of Energy

1. Solar energy passes through Earth's atmosphere and warms the ground.
2. Some energy is reflected back into space by the atmosphere.
3. Some energy is reflected back into space from the surface.
4. Earth's surface radiates some of its warmth into space.
5. The atmosphere traps some of the heat and stops it from escaping into space.

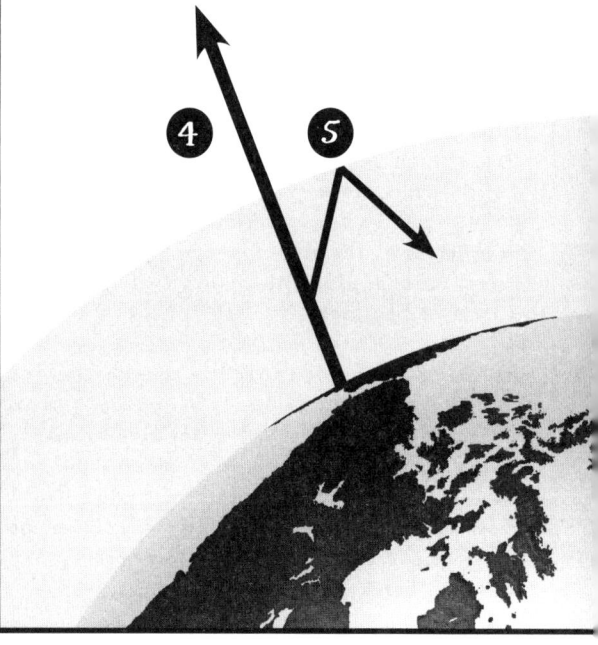

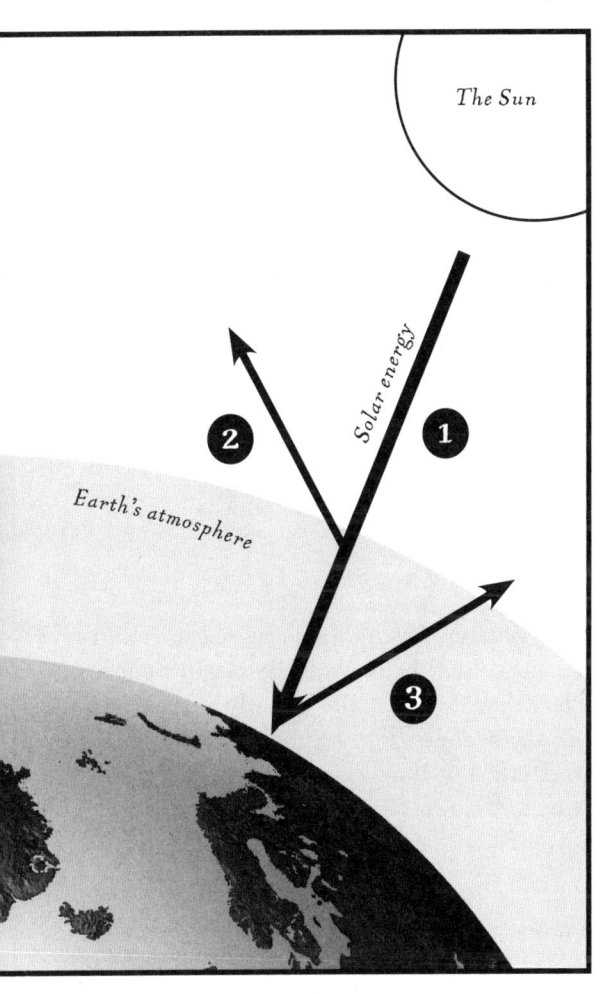

Here are the main gases in our atmosphere:

> Nitrogen: 78 per cent
> Oxygen: 21 per cent
> Argon: 0.93 per cent
> Carbon dioxide: 0.04 per cent
> Trace amounts of neon, helium, methane, krypton and hydrogen, plus water vapour.

A list of the greenhouses gases, with the most abundant at the top, is markedly different:

> Water vapour
> Carbon dioxide
> Methane
> Nitrous oxide
> Ozone

Water vapour, which fluctuates in quantity, bounces back heat from the Earth's surface.* The other GHGs make up only a small part of the atmosphere, but without them we'd freeze to death: the Earth is more than 30°C (48°F) warmer than it would be without them.

---

* Conversely, the upper white surfaces of clouds reflect solar radiation back into space – a process known as the 'albedo'. Ice has the same effect.

The carbon cycle

Although CO_2 gets a bad press, carbon is the stuff of life. There are about 65 billion metric tons of it on Earth, stored in rocks and fossil fuels, and roaming free in the atmosphere, the seas, soil, plants and animals – indeed, in our very DNA.

In what's known as the carbon cycle, this restless element is always on the move.

Mighty methane

Alessandro Volta (1745-1827), the Italian scientist who invented the electric battery, first discovered methane (CH_4) at Lake Maggiore in the 1770s.

Although it accounts for only 0.00017 per cent of the atmosphere, it's 20 times more powerful than carbon dioxide as a greenhouse gas. It comes from volcanoes, bacteria in the ground, rubbish dumps and farm animals, and it hangs around for some ten years before it combines with oxygen to form carbon dioxide and water.

1. Carbon is released into the atmosphere through decomposing organisms, the exhalation of animals and the burning of fossil fuels.

2. Plants absorb CO_2 and, through the process known as photosynthesis, convert it into carbohydrates necessary for their growth.

3. Animals feed on the plants, passing carbon up the food chain.

4. When the plants and animals die they're eaten by decomposers, which return their carbon to the atmosphere as $CO2$.

Vitally for us, photosynthesis produces a byproduct: oxygen. The exhalations of trees and other plants give us the air we need to breathe.

Sink or Swim

Left to its own devices, the carbon cycle proved pretty stable over the millions of years before human exploitation began pumping huge amounts of CO_2 into the atmosphere. Now that we've disturbed that balance, scientists are anxious to discover how far the Earth is able to cope with the overload.

A gasping mouse

Put a mouse inside a sealed bell jar with a burning candle and a mint plant and you can test how photosynthesis works.

No, we're not recommending it, but this is the method used by Joseph Priestley (1733–1804), the Yorkshireman credited with discovering oxygen* and the carbon cycle. The poor mouse lost consciousness (and the candle guttered and smoked) when shut in the jar by itself, but it survived much longer (and the candle flared) when it had the mint for company.

Priestley, who also invented soda water, was a controversial dissenting theologian and a political theorist as well as a chemist. After a mob burned down his home and his church in Birmingham ('the Priestley Riots') because of his religious beliefs and his support for the French Revolution, he fled first to London and then to Pennsylvania in the USA, where he spent the last ten years of his life.

* *He called it 'dephlogisticated air': the French nobleman and chemist Antoine Lavoisier (who also has claims to the discovery) gave it the name we know today.*

The question is whether some of the excess will find its way into the natural 'carbon sinks' which have hitherto done the job via the process known as carbon sequestration.

- **The soil beneath our feet is thought to hold three times as much carbon as is found in the atmosphere.**

- Plants take carbon from the air and, when they die, deposit some of it in the soil.

- CO_2 dissolves in water, and the oceans are a huge store of carbon.

Hot topic

We make no apologies for referring to **global warming** throughout this book – it is, after all, the most pressing problem of our times – but we should note that scientists use the term specifically to mean temperature rises caused by human activity since the start of the industrial revolution.

They distinguish it from **climate change**, which has natural as well as human causes and produces often complex side-effects.

LIVING IN A GREENHOUSE

Any answer to this question will depend upon the extent to which we reduce, maintain or increase our generation of greenhouse gases in the years immediately ahead of us.

Our green and pleasant troposphere is a place of delicate physical balances which we disturb at our peril – and the clear evidence of recent years is that we've already set in motion a catastrophe that will be difficult to reverse...

Deep sea journey

We're all familiar with the tides that ebb and surge on our beaches, but far from the naked eye there's a kind of powerful deep-sea conveyor belt which, driven by temperature and salinity, is constantly moving the oceans on a journey around the globe.

Warm water begins to cool down as it flows northwards from the tropics, and when it gets cold enough it becomes more dense and sinks. It then flows back towards the tropics along the bottom of the ocean and continues on its long journey around the world – rising to the surface again in the Indian Ocean and the Pacific Ocean.

The Great Ocean Conveyor* isn't in a hurry: a round trip takes about 1,600 years.

Britain and the Scandinavian countries are warmed by a current known as the North Atlantic Drift, which is part of the Gulf Stream that comes all the way from the Gulf of Mexico.

* *If you want to show off you can refer to it by its scientific name 'Thermohaline circulation'.*

And what if the conveyor belt should ever grind to a halt? This would disturb the balance of the climate, cutting off the warm water from the tropics. Ocean temperatures in the Arctic would fall, cooling the air above it, and the average temperature would drop by about 8°C (14.4°F).

The Great Ocean Conveyor

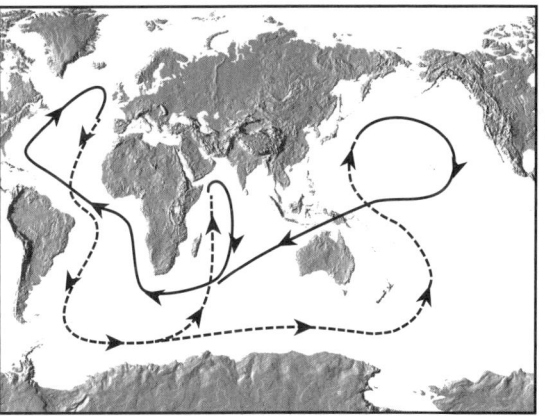

⁓ = Flow of hot water
--- = Flow of cold water

'We're sacrificing the lives of the next generation for our own. Not even for survival, but for comfort.'

CHAPTER TWO

WHAT ON EARTH HAVE WE DONE?

The hundred mourners who trudged up the lower slopes of a mountain in Iceland one Sunday in August 2019 and solemnly affixed a bronze memorial plaque to bare rock were taking part in a funeral with a difference: their dear departed was a glacier.

Back in 1890 the Ok glacier covered 16 square kilometres (6.2 square miles), but now all that remained was a small, static patch of ice. It was the first of the country's moving ice sheets to die, and those who gathered to mark its passing knew that hundreds of others would eventually meet the same end.

The plaque has the heading 'A letter to the future', and its message is starkly simple:

> Ok is the first Icelandic glacier to lose its status as a glacier. In the next 200 years all our glaciers are expected to follow the same path. This monument is to acknowledge that we know what is happening and what needs to be done. Only you know if we did it.
>
> August 2019
> 415 ppm CO_2

The anthropologist Cymene Howe from Rice University in Texas was one of the event's organisers. 'By memorialising a fallen glacier,' he said, 'we want to emphasise what is being lost – or dying – the world over, and also draw attention to the fact that this is something that humans have "accomplished", although it isn't something we should be proud of.'

The author and environmentalist Andri Snær Magnason, who devised the plaque's wording, spoke of 'an unbearable existential dilemma': 'We're sacrificing the lives of the next generation for our own. Not even for survival, but for comfort.'

WHAT ON EARTH HAVE WE DONE?

Iceland is losing about 11bn tonnes of ice each year, but it's not alone in its unfreezing.

- In the Swiss Alps residents wrap the Rhône Glacier in white canvas blankets during the summer months in order to reflect sunlight away from it – a practice also followed in mountainous areas of Italy and Germany.

- In September 2019, echoing the Ok 'funeral', the Sankt Gallen canton marked the first death of a Swiss ice sheet, the Pizol Glacier.

- In Sweden, Kebnekaise lost its status as the country's largest mountain in 2019 because the glacier on its summit had shrunk.

In proportion

In 2010 Iceland's Eyjafjallajökull volcano erupted, the ash clouds so dense that air travel across Europe was cancelled for several days. But a comparison: its CO_2 emissions amounted to 150 tonnes a day, whereas human activity worldwide produces almost 100m tonnes a day – or all of 600 Eyjafjallajökulls.

- Glacier National Park in Montana, USA, had 150 glaciers when President Taft created it in 1910, but now there are fewer than 30 – and they're dwindling.

- Satellite images of the remote Himalayas show that a quarter of the area's glacier ice has melted in the last 40 years, the runoff forming glacial lakes which are liable to flood, with disastrous results.

Grolar bears

Who would ever have imagined that global warming would produce a hybrid animal?

Grizzly bears live on land, while polar bears live on ice, so normally the twain never meet. A warmer climate, however, allows the grizzlies to live further north than before, while melting sea ice forces polar bears to move further south and onto land.

The two species are now meeting and doing what bears do. The result is the grolar bear, a new hybrid. Its existence was confirmed by DNA analysis of a strange-looking bear shot in the Canadian Arctic in 2006.

- Glaciers in the Peruvian Andes, home to 70 per cent of the world's tropical glaciers, have retreated by 40 per cent since the 1970s, and the famous Pastoruri Glacier, long a tourist attraction, has now become part of a climate change tour.

Deadly secrets

When unaccustomed heat thaws the frozen tundra that we know as permafrost it can have startling, sometimes dangerous, consequences – and perhaps no more dramatically than in the chilly wastes of Siberia.

Here the softened earth has parted to reveal ancient corpses, including woolly mammoths which last walked the Earth thousands of years ago. But scientists fear that the newly liberated soil and water may also contain dormant viruses and bacteria that can survive in human and animal remains for hundreds of years – and when, in 2016, a young boy died in an anthrax outbreak among nomadic reindeer herders in Russia's Arctic Circle, a rise in temperature was thought to have released the deadly spores into the modern world.

An immediate danger of thawing permafrost is the cracking of fuel pipelines and the collapse of buildings as their foundations give way. In the Russian city of Norilsk, 290km (180 miles) above the Arctic Circle, more than half the houses have become 'slow motion wrecks'.

A less obvious danger is the slow release of carbon stored in the tundra, much of it that powerful global warming constituent, methane.

'The Arctic acts like a freezer,' the climate expert Joseph Romm has written, 'a very large carbon freezer, and the decomposition rate is very low, or at least it has been. We are in the process of leaving the freezer door wide open. The tundra is being transformed from a long-term carbon locker to a short-term carbon unlocker.'

Swelling seas

Collapsing (or 'calving') icebergs have become among the most spectacular symbols of climate change, and the melting ice has only one place to go. Global warming is already raising sea levels, and there is worse to come.

Earth Overshoot Day

We're steadily using up more of the Earth's resources than it can regenerate, and the date each year by which humanity has overspent its ecological budget is known as Earth Overshoot Day.

It is, of course, a rough estimate based on a huge raft of figures, but the results are alarming. Back in 1987 we went into the unsustainable red on October 23rd, whereas by 2019 the tipping point was July 29th.

This is a global figure, and many individual countries do far worse. Here are the top ten guilty parties in 2019:

Qatar February 11
Luxembourg February 16
United Arab Emirates March 8
Bahrain March 10
Kuwait March 11
Trinidad and Tobago March 13
USA March 15
Canada March 18
Mongolia March 19
Bermuda March 21

And the UK? In 49th place, May 18.

Saving London

Completed in the early 1980s to protect the low-lying capital from storm surges and exceptionally high tides, the Thames Barrier straddles a 520m (1,700ft) wide stretch of the river east of the Isle of Dogs. Its rotating cylindrical gates were the brainchild of the engineer Charles Draper – based on the design of the taps on his gas cooker.

By the spring of 2019 the gates had been raised on 184 occasions. In view of climate change, there have been calls for a new, stronger defence further downstream, but the Environment Agency is confident that the barrier won't need replacing before 2070.

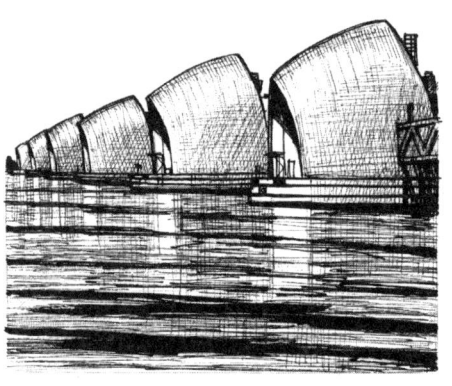

WHAT ON EARTH HAVE WE DONE?

In 2008, 414 square kilometres (160 square miles) of the vast Wilkins Ice Shelf broke away from the Antarctic coast; the Western Arctic Ice Sheet is currently losing 150 cubic kilometres (36 cubic miles) of mass every year; while a heatwave in July 2019 caused Greenland's ice to melt at a rate scientists weren't expecting for another fifty years.

Some 55 billion tons of it disappeared in five days – enough, it was calculated, to cover the entire state of Florida in almost 12.7cm (5in) of water.

Coastal communities are threatened across the world.

- Vietnam, Pakistan, Cyprus, Morocco and Tunisia are among countries which have suffered saltwater contamination of their drinking water supplies.

- In 2019 Indonesia's president announced plans to move the nation's capital from Jakarta on the island of Java – for several reasons, but not least because the land is sinking at the rate of 25cm (10in) a year.

- The Republic of Kiribati, a nation of islands and atolls in the central Pacific which gained its independence from the UK in 1979, has already lost two of its islands to the rising sea. In 2018 President Taneti Maamau declared scepticism about man-made climate change and announced plans for new luxury resorts, but the former leader, Anote Tong, delivered a sharp retort:

'It is time for the world to wake up and understand,' he said. 'We are all Kiribati.'

Underwater signals

A similar controversy rages in the Maldives, although the climate aware former President, Mohamed Nasheed, didn't see it coming.

In October 2009 he held the world's first underwater cabinet meeting, a stunt in which he and his fellow politicians donned scuba gear and sat at desks on the seabed, communicating with each other by hand signals.

Nasheed was so alarmed by future swamping that he considered buying a new homeland in India, Sri Lanka or Australia in case everyone had to leave the islands.

WHAT ON EARTH HAVE WE DONE?

The future of the Maldives does, indeed, seem parlous. It has 1,192 islands, of which 200 are inhabited by about 400,000 people. Not only is the average ground level just 1.5m (5ft) above sea level, but its highest point, at 2.3m (7½ft) above sea level, is the lowest of any country in the world, and any significant rise in the lapping waters would make the whole nation uninhabitable.

But what do we find? In 2017 the country's new president, Abdulla Yameen, decided that the future lay, instead, in depopulating the smaller islands, geo-engineering artificial ones and attracting millions of tourists by creating fifty glistening new resorts.

Although an IPCC report had forecast that 75 per cent of the Maldives would be under water by 2100, the director of the government's Marine Research Centre, Shiham Adam, was having none of it.

'We have immediate needs,' he told journalists. 'Development must go on. Jobs are needed. We have the same aspirations as people in the US or Europe.'

Stricken reefs

Coral reefs are among nature's wonders, and they're now in serious danger from the warming of the oceans.

They may have the appearance of plants, but corals are marine animals, and their colour largely derives from algae which live in a symbiotic relationship with them. In the 'bleachings' of 2017 and 2018, around half of the corals in the Great Barrier Reef which runs for 2,300km (1,400 miles) down Australia's north-east coast turned a ghastly white and died as the algae* succumbed to the higher temperatures.

All the reefs in 29 World Heritage Sites are under threat – and don't imagine that it's only the prettiness we should worry about. They harbour the greatest biodiversity of any ecosystem on the planet, accounting for a quarter of all marine life on the planet.

Algae in the oceans give off more oxygen than all land plants combined.

For those who like to count the cost financially, the journal *Global Environmental Change* has calculated the social, cultural and economic value of coral reefs to be $1 trillion, while the World Wildlife Fund has forecast that the climate-related loss of reef ecosystem services will cost $500 billion or more per year by 2100. It's reckoned that they support over 500 million people worldwide, most of them in poor countries.

Acid test

Those oceans, expanding with the heat, present us with another problem – one that's been described as 'the evil twin of climate change'. They absorb roughly a quarter of the fossil fuels we send into the air, and the CO_2 dissolves in them to form carbonic acid.

The oceans have acidified by almost 30 per cent since the beginning of the industrial revolution, and a study in 2010 found that they're doing so ten times faster today than at any period since 55 million years ago when there was a mass extinction of marine species.

Acidification is affecting the ability of many marine organisms, corals included, to form their shells, and in the most serious conditions it can actually dissolve them.

Lobsters, crabs and octopuses are at risk, and the major oyster businesses off the Californian coast are among those threatened. Tessa Hill, a marine scientist at the University of California who investigated conditions in the 19km (12 mile) Tomales Bay estuary, foresaw 'a blanket of corrosive water' off the coast by 2050.

All this and plastics too

The scourge of plastics in the environment isn't, of course, a climate change issue, but it's another example of the damage we've done to the planet.

It's estimated that more than 300,000 marine animals are harmed or killed by plastic every year – many ingesting the stuff, which clogs up their digestive systems.

The Great Pacific Garbage Patch is the largest accumulation of plastic in the world.

WHAT ON EARTH HAVE WE DONE?

A burning shame

Human greed and blindness is nowhere on more terrifying display than in the destruction of our vital tree cover. A September 2019 report by the think tank Climate Focus revealed that the world was losing an area of forest the size of the United Kingdom every year, most of it tropical rainforest, with grim effects on both the climate and wildlife.

In the days before the report was published the media had published dramatic photographs of huge tracts of the Amazonian rainforest burning in Brazil, a conflagration caused not by a random act of nature but by loggers and developers creating land for farming.

This biological carnage is a double whammy: the smoke from the fires adds to concentrations of CO_2 in the atmosphere, while the trees' free gift of oxygen is gone for ever – traded for piles of lifeless ashes.

Alarmingly, the rate of tree loss had risen by 43 per cent since the New York declaration signed at the UN in 2014 requiring countries to halve deforestation by 2020.

'We don't need more important guys standing up making pledges,' said the think tank's co-founder Charlotte Streck when presenting the report. 'We need to go beyond declarations. Implementation is complicated, but it's what we need... It can take centuries for forests to recover their full carbon-absorbing and weather-regulating capabilities.'

- In Latin America, south-east Asia and Africa the annual rate of tree cover loss had risen substantially between 2014 and 2018 and the most valuable and irreplaceable tropical primary forests had been cut down at the rate of 4.3 hectares (10.6 acres) a year.

- Many governments offered subsidies to agriculture, so providing incentives for deforestation, and there were few investments in keeping forests healthy.

- On the brighter side, Indonesia's destruction of primary forest (chiefly to create virgin land for palm oil plantations) had slowed by almost a third because companies and the government had come under pressure from consumers and aid donors.

In Brazil, meanwhile, the climate change denying president Jair Bolsonaro (who proudly called himself Captain Chainsaw) adopted an aggressive stance in favour of agribusiness and against both the indigenous population living in the rainforest and foreign critics who accused him of threatening the future of the planet.

Even before taking office in January 2019 he had rescinded Brazil's offer to host the UN climate change conference that year.

Vicious cycle

The forest fires in Brazil came at the end of the wet season, when the trees should have been hardest to burn, and this led climate scientists to warn of 'a feedback loop'.

Global warming dries out the trees, making them more flammable. When they are torched, the CO_2 intensifies the heat in the atmosphere yet further, thus drying out the remaining trees and making future burning easier to achieve, naturally or otherwise.

Between January and September 2019 the National Institute for Space Research documented almost 40,000 fires in the Amazon rainforest, an increase of 77 per cent on the previous year – and in July an area the size of Manhattan was being lost every day.

Carbon footprints

Yes, we're all guilty. You don't have to drive a car or fly in a plane to add carbon dioxide or methane to the atmosphere – even breathing out does it!

Your carbon footprint is the amount of emissions you're responsible for, not only directly but indirectly when (for instance) you buy food which has required the burning of fossil fuels in order to deliver it to your supermarket shelves.

It's an inexact calculation, of course, but it compares what individuals, organisations, events or products demand of the planet compared with what it's able to renew.

If you can bear to learn the harsh truth, there are several free carbon footprint calculators online.

Bolsonaro's response was to blame the fires on environmentalists and then to describe offers of help from other countries as 'colonialism'.

Burping cows

It follows from the above that farming isn't always a good thing. Destroying rain forests to grow crops or herd cattle is a short-term gain producing a serious long-term deficit.

Carbon may be the number one enemy, but the gases arising from nitrogen in fertilisers and livestock effluent are some three hundred times more effective than CO_2 in trapping heat in the atmosphere.

And then there's the methane produced in huge quantities in the stomachs (four each) of cud-chewing cows.* Most of it is burped out, some emerges at the rear end – and each of them produces around 200kg (440lbs) a year.

---

* *Rice growing is another culprit. Flooded paddy fields stop oxygen penetrating the soil – ideal conditions for the methane-emitting bacteria responsible for around 1.5 per cent of total greenhouse emissions.*

Researchers may have found some answers to this problem, though. A study published in the journal *Science Advances* in 2019 showed that cows vary in the amount of methane they produce according to the particular microbes they inherit in their guts. Selective breeding may be able to engineer a low-methane 'microbiome' to produce more environmentally friendly cows.

Or perhaps a dietary solution will suffice instead. It seems that low doses of the Australian red seaweed *Asparagopsis taxiformis* virtually eliminate methane. The only snag is that the seaweed will have to be produced in industrial quantities.

'Natural' disasters

Although no individual weather event can be directly attributed to man-made climate change, the ever increasing and intensifying occurrence of what would once have been described as natural disasters, or even Acts of God, is an alarming consequence of our heating the atmosphere to a level the human race has never experienced before.

WHAT ON EARTH HAVE WE DONE?

Two decades into the 21st century the world is afflicted by a relentless succession of calamities that threaten the lives of millions:

- **Supercharged storm surges and tsunamis which now threaten a quarter of the world's population.**

Storm proof

The ingenious Maeslantkering storm surge barrier protecting the port of Rotterdam is an example of what nations at risk of flooding will have to create – if they can afford it.

Controlled by a supercomputer, the barrier comprises two huge, hollow steel gates which are stored in dry docks until a surge warning is issued. They then float out across the mouth of the Nieuwe Waterweg ship canal, meet in the middle, fill with water and sink, so sealing off the port from the sea.

The barrier, opened in 1997, was designed to withstand a one-in-10,000 years storm event, although this will have to be adjusted to allow for climate change.

- Hurricanes of increasing violence.

- Rising sea levels, swamping low lying communities.

- Unbearable heatwaves and wildfires.

- Prolonged and intense droughts.

- Torrential rainfall, with severe flooding.

Deadly blasts

Hurricanes suck up warm tropical water as fuel, and their ferocity is growing stronger. The highest category, 5 on the Saffir-Simpson scale, represents a force of 252 kph (157mph) or more, but Dorian - which devastated the Bahamas in August and September 2019 - swept ashore at all of 295kph (185mph), a record for an Atlantic hurricane at landfall.

During the same year the tropical cyclone Idai killed more than 1,300 people in Mozambique, Malawi, Zimbabwe and Madagascar, while tropical typhoon Lekima in the Western Pacific Ocean struck China, causing damage estimated at more than $9 billion.

WHAT ON EARTH HAVE WE DONE?

> 'The current rate of increase in global warming is roughly the same as detonating 400,000 Hiroshima bombs per day, 365 days per year.'
>
> *Joseph Romm*

All the while that these catastrophes have been playing out around us, we have continued to pump fossil fuels into the atmosphere in increasing quantities and, in consequence, the global temperature has steadily risen.

Climate change isn't something waiting to happen, but a process already well under way, and its fateful onward progress can be slowed only by the people who set it in train.

If we fail, then forget the alarming events outlined in this chapter. You ain't seen nothing yet...

Pesky children

Every three to seven years a mysterious switching of winds and ocean currents in the Pacific brings weather changes that are often disastrous. It's known as El Niño* (Spanish for a young boy – and in particular the Christ child, as it tends to start around Christmas), and its effects can last for more than a year.

The trade winds normally blow from east to west across the tropical part of the Pacific, from Ecuador to Indonesia, blowing warm surface water before them and drawing deep cold water to the surface of South America. This cools the air above it, producing the region's dry weather, while Indonesia receives a heavy dousing of warm rain.

During an El Niño event this pattern is reversed, with Indonesia parched and Peru and Ecuador experiencing floods. With no upwelling of the usual cold nutrient-rich water off the South American coast, fishing fleets are grounded.

*Although scientists now prefer to call it ENSO, for El Niño–Southern Oscillation. Get a life, chaps!

There are knock-on effects across the world, including the strength and timing of the Asian monsoon and the number, intensity and direction of hurricanes and cyclones.

During 1997-98, the worst disruption so far recorded, El Niño caused droughts in the southern USA, East Africa, northern India, north-east Brazil and Australia. Forest fires raged in Indonesia, and there was torrential rainfall with flooding in parts of South America, Sri Lanka and East Central Africa.

El Niño has a little sister, La Niña, who follows hard on his heels – often lingering for a full year. On these occasions the eastern Pacific is cooled well below normal temperatures by an upwelling of cold water, while rainfall increases over the western Pacific.

Britain is too far away to be seriously affected, although El Niño has on occasion been praised for warm summers and La Niña blamed for chill winters. The fact is that the scientists are still researching these curious phenomena, and the connection between 'children' events and global warming isn't yet properly understood.

'this is the moment and we must act now'

CHAPTER THREE

SURVIVING THE CENTURY

In September 2019 the UK's former chief scientific adviser, Professor Sir David King, told the BBC that recent heatwaves and other extreme weather events had so far exceeded what climatologists had forecast that it was 'appropriate to be scared'.

Scientists don't often use emotive language, but Professor Joanna Haigh, a physicist from Imperial College London, was quick to agree with him. 'Dr King is right to be scared,' she said. 'I'm scared too. We do the analysis, we think what's going to happen, then publish in a very scientific way. Then we have a human response to it – and it *is* scary.'

Sir David, having gazed into his global warming crystal ball, presented the UK government with a challenge he must have guessed they would duck: to advance its climate targets by ten years, cutting greenhouse gas emissions to around zero by 2040 rather than 2050.

Asking the questions

Scientists are professional sceptics. Knowing that certainty is rarely possible, they test their results over and over again, always looking for a rogue miscalculation which might invalidate their findings.

So many of them have spent so many hours tussling with climate change trends and statistics that many of the basic elements are now well understood, but none of them would bet the house on precise forecasts for the years ahead.

That said, the projections in this chapter are far from wild guesses, and here's a sobering thought: in general the acceleration of climate change and the severity of its consequences have been greater than forecasters predicted.

All their tomorrows

Most children born today have a realistic expectation of living until the end of the 21st century, but what kind of a world will they know? How many parts of the Earth will be habitable in 2100, and how much wildlife will be left to share it with?

The table overleaf gives estimates of the damage caused to our environment depending on the rate at which we keep heating it – *and if you bookmark only one page in our little guide, this should be it.*

Stopping the advance of global warming in its tracks is impossible. A 2008 study by *Nature Geoscience* found that we've been releasing CO_2 into the atmosphere 14,000 times faster than natural processes managed over 600,000 years, and those concentrations would take centuries to disperse even if we stopped using fossil fuels overnight.

The race is on to mitigate the damage by keeping the global temperature rise to under 2°C (3.6°F) – and that won't be easy.

Our heating atmosphere

Projected effects of rises in temperature above pre-industrial levels

1°–2°C (1.8°–3.6°F)

Major impacts on vulnerable ecosystems – polar regions, wetlands and cloud forests.

Increases in extreme weather events.

Spread of infectious diseases.

2°–3°C (3.6°F–5.4°F)

Major loss of coral reefs.

Serious impacts on agriculture, water resources and health.

All ecosystems and species under severe stress.

Significant rise in extreme weather events.

Terrestrial carbon sink expels CO_2, so accelerating climate change.

3°–4°C (5.4°–7°F)

Major species extinction.

Lack of food and clean water, with serious impact on health.

Environmental depletion leading to mass migrations.

Failure of oceanic carbon sink, accelerating CO_2 accumulation in the atmosphere.

Extreme weather events ten times more common than in 2010.

4°C (7°F)

Melting of Western and Antarctic and Greenland ice sheets with huge rises in global sea levels.

Drought and desertification.

Further mass migrations as resources are depleted, leading to possible civil unrest.

Huge loss of biodiversity.

4°–5°C (7°–9°F)

Water shortages suffered by more than 3 billion people.

A fifth of the world's population affected by flooding.

Collapse of food production, leading to widespread famine.

A significant rise in human mortality because of malnutrition, starvation, disease, flooding and extreme weather events.

5°–6°C (9°F–11°F) and above

Human life practically insupportable.

Lagging behind

In October 2018 the IPCC re-emphasised the need to restrict the increase in global warming to 1.5°C (2.7°F) and warned that the world had only a dozen years to achieve that goal.

Without 'urgent and unprecedented changes', declared a landmark report, millions would be condemned to drought, floods, heat and poverty, and there would be a devastating collapse of wildlife habitats.

'It's a line in the sand,' said Debra Roberts, a co-chair of the IPCC working group, 'and what it says to our species is that this is the moment and we must act now. This is the largest clarion bell from the science community and I hope it mobilises people and dents the mood of complacency.'

Months later, in June 2019, Climate Action Tracker (a project run by three research organisations) issued a report on the climate change commitments of 32 countries which, between them, account for 80 per cent of man-made greenhouse gases.

SURVIVING THE CENTURY

The figures were depressing, seeming to render the chances of meeting that 1.5°C target slim to non-existent.

With the clock loudly ticking, most of the major polluters were making few, if any, efforts to clean up their act.

CRITICALLY INSUFFICIENT: 4°C (7°F) PLUS
Russian Federation, Saudi Arabia, Turkey, Ukraine, USA.

HIGHLY INSUFFICIENT: 4°C (7°F)
Argentina, Chile, China, Indonesia, Japan, Singapore, South Africa, South Korea, UAE.

INSUFFICIENT: 3°C (5.4°F)
Australia, Brazil, Canada, EU, Kazakhstan, Mexico, New Zealand, Norway, Peru, Switzerland.

2°C (3.6°F)
Bhutan, Costa Rica, Ethiopia, India, Philippines.

1.5°C (2.7°F) PARIS AGREEMENT COMPATIBLE
Morocco, The Gambia.

Amid a growing awareness of pending disaster, the United Nations called a Climate Action Summit in New York for September 2019, its secretary general, António Guterres, urging delegates before the event, 'Don't bring a brilliant speech – bring a plan.'

The Doomsday Clock

In 1947 the *Bulletin of the Atomic Scientists* in Chicago devised its Doomsday Clock, registering an assessment of threats to humanity from unchecked scientific and technical advances.

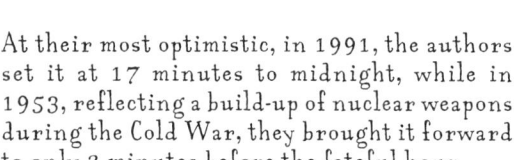

At their most optimistic, in 1991, the authors set it at 17 minutes to midnight, while in 1953, reflecting a build-up of nuclear weapons during the Cold War, they brought it forward to only 2 minutes before the fateful hour.

It swung many times after that, but returned to the 2-minute mark in 2018 (and stayed there in 2019) as climate change became the most ominous threat to the planet.

As the summit opened, delegates were greeted by a devastating United in Science report coordinated by the World Meteorological Organization. It warned that even if countries honoured their current plans, the average rise in global temperatures would rise by between 2.9° (5.2°F) and 3.4°C (6.1°F) by the end of the century.

And its solution was drastic: existing commitments to cut greenhouse emissions would have to be at least tripled if temperatures were to be checked at 2°C (3.6°F) above pre-industrial levels and increased fivefold to meet the 2015 Paris target of 1.5° (2.7°F).

Speeches were denied to countries bent on building and financing new fossil plants or found to be neglecting their earlier promises, among them the US, Japan, South Africa, Saudi Arabia and Brazil – and Donald Trump made only a fleeting appearance.

'How many climate records does it take,' asked Pep Canadell, a contributing author to the report, 'to accept the unprecedented nature of what we are living and to act upon it?'

Of course there were both brave speeches and promises of action, but only the brightest of optimists came away denying that serious episodes of self-harm await us as the century unfolds.

The question is, how serious will they be?

The great melt

We've already seen that the melting of glaciers and the great ice sheets, together with the expansion of warming oceanic water, has produced a rise in sea levels across the globe.

It's been calculated that if the Greenland and West Antarctic ice sheets collapsed entirely (and they've proved to be surprisingly vulnerable to global warming) they would raise sea levels by at least 4.5–6m (15–20ft).

None of the experts, comfortingly, is predicting anything quite so extreme, but many of them believe that a rise of 1.8m (6ft) by 2100 is highly likely, and that would bring misery to millions of people living in low-lying areas around the world.

The cruel sea

The 20 cities predicted to be most vulnerable to flooding by 2050, published in the magazine *Nature Climate Change*.

1. Guangzhou, China
2. Mumbai, India
3. Kolkata, India
4. Guayaquil, Ecuador
5. Shenzhen, China
6. Miami, USA
7. Tianjin, China
8. New York/New Jersey, USA
9. Ho Chi Minh City, Vietnam
10. New Orleans, USA
11. Jakarta, Indonesia
12. Abidjan, Ivory Coast
13. Chennai, India
14. Surat, India
15. Zhanjiang, China
16. Tampa, Florida, USA
17. Boston, Massachusetts, USA
18. Bangkok, Thailand
19. Xiamen, China
20. Nagoya, Japan

Others are even more pessimistic. In 2015 a team led by James Hansen, a leading American climatologist, warned that failure to curb the advance of CO_2 might lead to a rise of 3m (10ft) in sea levels by 2001.

Two years later a National Climate Assessment report approved by Congress suggested not only that a rise exceeding 2.4m (8ft) by the end of the century was 'physically possible', but that current trends suggested a rise of 3m (10ft) each decade after 2050 and a doubling of that rate after 2100.

'I can't envisage south-eastern Florida having many people at the end of this century,' Harold Wanless, chair of the University of Miami's geological sciences department, said in 2013. 'Miami, as we know it today, is doomed. It's not a question of if; it's a question of when.'

By 2050 more than 570 coastal cities are expected to be in trouble, most of them in Asia and North America – although we can all think of places closer to home that would have trouble withstanding such a prodigious swelling of the tides.

Bird, beast and flower

The world's wildlife has been disappearing at an unprecedented and accelerating rate.

- **The biomass (or total weight) of wild mammals on land and in the sea has fallen by 82 per cent.**

- **Well over half of forest wildlife has been lost since 1970, the greatest depletion occurring in rain forests.**

- **More than 28,000 species on Earth are threatened with extinction, including 40 per cent of amphibians, 25 per cent of mammals and 14 per cent of birds.**

These shocking losses are by no means all the result of climate change. Overfishing of the oceans and the destruction of habitats in order to turn them over to agriculture are prime examples of human harm to biodiversity without a direct carbon connection.

The undeniable fact, however, is that global warming and its knock-on effects are putting a massive strain on an ecosystem already at breaking point.

Here's a World Wildlife Fund list of ten animals most at risk from climate change this century, along with their Red List* status:

- **Polar Bear**: vulnerable
- **Snow Leopard**: endangered
- **Giant Panda**: endangered
- **Tiger** (*Panthera tigris*): endangered
- **Monarch Butterfly**: migration endangered
- **Green Sea Turtle**: endangered
- **African Elephant**: vulnerable
- **Mountain Gorilla**: endangered
- **Asian Elephant**: endangered
- **Cheetah**: vulnerable

The virtue of lists like these is that, in alerting the public to the charismatic creatures we'd all be sad to lose, they encourage campaigns to save them.

What they disguise, on the other hand, is the value of far lesser forms of life (in popstar terms, that is) which play a vital role in maintaining the essential balance of nature.

---

* Compiled by the International Union for Conservation of Nature (IUCN), the world's largest environmental network.

To put it bluntly, we'd shed a tear if the Panda disappeared, but its extinction would have no noticeable environmental effect – whereas if we lost the algae from the oceans and insects from the land the impact of this loss of diversity would soon become frighteningly apparent.

As it is, a million species (half of them animals and plants, the other half insects) are now in danger of total obliteration.

Hungry puffins

A shrinking in the puffin population off England's north-east coast demonstrates the effect of climate change on the food chain – and the role of a zooplankton no bigger than two grains of sugar.

The puffins feed on sand eels, greedily (and rather comically) gathering large quantities of them in their beaks. In recent years, however, the eels have been in short supply. Why? Well, they in turn make a meal of the plankton *Calanus finmarchicus*, and as the Atlantic has grown steadily more acidic so numbers of their tiny prey have dwindled.

Extinction Rebellion

The campaigning group Extinction Rebellion (or XR) was formed in the UK in 2018, backed by around a hundred academics, with a demand for speedy governmental action on biodiversity loss and climate breakdown. Among its many stunts was the closing of five bridges on the Thames, pouring buckets of fake blood on the road outside Downing Street to represent the threatened lives of children and occupying prominent sites in central London, including an area around Parliament Square. Hundreds were arrested.

Its demands included the establishment of a citizens' assembly on 'climate and ecological justice' and a zero-carbon target for the UK by 2025 – though critics otherwise sympathetic with its aims were scathing about the proposed timescale, saying that it could be achieved only by grounding all aircraft flights, taking 38 million cars off the roads and disconnecting 26 million gas boilers within six years.

With a stylised hourglass as its symbol, the organisation and its disruptions soon spread to other countries, including Australia and the USA.

According to Robert Watson, a former chair of the UK's intergovernmental biodiversity service, 'the situation has become more and more dire'.

All the international biodiversity assessments he had headed over the years had repeated the same message: 'We are destroying it at an alarming rate. Each time we have called for action, only to be largely ignored.'

Life's little helpers

Insects may buzz and bite, but they're busy pollinators of our crops and other plants, and they provide food for many birds, reptiles, amphibians and fish.

In 2019 the journal *Biological Conservation* published a global review revealing that 40 per cent of insect species were declining, while a third were endangered – a rate eight times faster than that for mammals, birds and reptiles. If this couldn't be stopped, it would have 'catastrophic consequences for both the planet's ecosystem and for the survival of mankind'.

One of its co-authors, Francisco Sánchez-Bayo from the University of Sydney, Australia, told the *Guardian* newspaper that there had been a 'shocking' loss of 2.5 per cent every year over 30 years.

'It's very rapid,' he said. 'In ten years you will have a quarter less, in fifty years only half left and in a hundred years you will have none.'

Deadly skeeters

There are, though, winners as well as losers when the climate changes. Disease-carrying mosquitoes love warm, moist conditions, and they're already spreading to areas beyond their normal range.

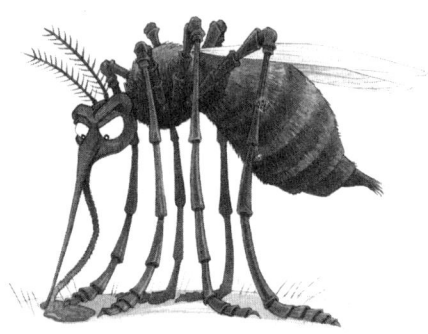

Experts have predicted that from 2030 onwards, at predicted levels of heat rise, the UK will be vulnerable to several 'exotic' mosquito-borne* diseases associated with far away places. Some have already made a first appearance in parts of Europe.

This list is not for the faint-hearted:

Malaria
Dengue fever
Chikungunya
West Nile virus
Mediterranean spotted fever
Zika virus
Leishmaniasis

The link between ticks and climate change isn't clear-cut, but they're certainly thriving – and in feeding on blood from animals and humans they transmit infections such as Lyme disease and Crimean-Congo haemorrhagic fever.

---

* *What they'll just love are the water butts which, in response to global warming and water shortages, have become a common feature of urban gardens: stagnant water is ideal for mosquitoes at egg-laying time.*

The World Health Organisation (WHO) has predicted that climate change will cause getting on for 250,000 *additional* deaths across the globe between 2030 and 2050. This is how they break the figure down:

- **95,000 Children undernutrition**
- **60,000 Malaria**
- **48,000 Diarrhoea**
- **38,000 Heat stress**

And a sobering note – the developed world creates most of the climate problems, whereas the developing world bears the brunt of them.

Downing tools

A little discussed aspect of a heating climate is the effect it will have on productivity, especially in agriculture and other outdoor industries.

The National Oceanic and Atmospheric Administration (NOAA) produced a study in 2013 which forecast a doubling of 'heat-stress related labor capacity losses' by 2050.

'If we stay near our current greenhouse gas emissions pathway, then we face a potential 50 per cent drop in labor capacity in peak (summer) months by century's end.'

In financial terms, this would mean a cost to society greater than from all other effects of climate change added together.

'According to our projections,' said Noah Diffenbaugh, lead author of a Stanford University study in the US, 'by the middle of this century even the coolest summers will be hotter than the hottest summers of the past 50 years.'

As for indoor work, a Japanese professor whose research was covered by the *New York Times* in 2012 calculated that 'every degree rise in temperature above 25°C (77°F) resulted in a 2 per cent drop in productivity'.

And heat isn't the only problem. Higher levels of CO_2 in offices have been shown to affect workers' decision-making abilities – as well, of course, as having a detrimental effect on their general health.

Armageddon

Those who bring bad tidings to the public are routinely accused of parading 'fake news' or launching 'project fear', but far too many scientists have posted the results of their detailed researches for there to be any doubt about the danger to our planet if governments fail to act swiftly against the heating of the atmosphere.

The fact that they have responded sluggishly to the crisis so far makes it inevitable that some of the horrors already foreseen in this book will come to pass.

What some critics fear (and there's a large and ever growing literature on the subject) is a breakdown of civil order across the globe, with desperate groups or nations driven to armed conflict.

For member countries of the European Union immigration from war-torn and economically deprived areas to the south has become a contentious problem that its governments have as yet been unable to solve. Populist leaders talk about 'swarms' of incomers.

But just consider the far greater stresses and strains that will be inflicted on the vulnerable – and will then impinge upon their more fortunate neighbours – as temperatures rise to cataclysmic levels:

- **Rising seas will swamp the homes of millions and ruin their supplies of fresh river water, driving them to seek new territory inland.**

- **In many areas the land itself will already have become unproductive, baked to an untillable hardness, producing dust bowl conditions and triggering prolonged droughts.**

Where will the dispossessed go? Who will be willing to accommodate such large numbers of refugees? What are the chances of a peaceful resolution to such a crisis?

In 2009 the UK government's chief scientist, Professor John Beddington, warned that, as early as 2030, 'a "perfect storm" of food shortages, scarce water and insufficient energy resources threatens to unleash public unrest, cross-border conflicts and mass migration as people flee from the worst-affected regions.'

Drought and a death cult

The rise of the murderous terrorist group known variously as ISIS and Islamic State had several causes, but climate change played a vital role in sparking Syria's civil war.

What's been described as 'the worst long-term drought and most severe set of crop failures since agricultural civilisations began in the Fertile Crescent' ruined the livelihood of 800,000 people and condemned many more to poverty in the five years to 2016.

A study published in *Proceedings of the National Academy of Sciences* found that global warming had made the Syrian drought two to three times more likely.

'The poor and the displaced fled to the cities,' it read, 'where government mismanagement and other factors created unrest that exploded in spring 2011.'

'While we're not saying the drought *caused* the war,' explained its lead author, Dr Colin Kelley, 'we *are* saying that it certainly contributed to other factors – agricultural collapse and mass migration among them – that caused the uprising.'

None of this is written in stone, of course. There is – just – time for governments to take what needs to be unprecedented drastic action to curb the relentless gushing of CO_2 into the atmosphere.

But to date most of the 'virtue signalling' about climate change has been reminiscent of the Emperor's New Clothes – and it's taken the eyes of a child to expose the sham for what it is...

Pointing the finger

A devastating report in October 2019 revealed that twenty fossil fuel companies had between them created more than a third of all greenhouse gas emissions in the modern era.

Seven-eighths of these emissions were from the use of their products in the way of petrol, jet fuel, natural gas and coal, the rest coming from extracting, refining and delivering the finished fuels.

The analysis by Richard Heede of the Climate Accountability Institute in the US also accused the producers of investing millions in 'climate denial and obfuscation' in order to delay legislative action which would affect their income.

The largest five oil and gas companies spent $200m a year on lobbying.

'The pursuit of next quarter's profits,' he said, 'must be shifted to embrace climate stewardship or our legacy will be a bereft planet and a broken moral compass.'

Twelve of the top 20 companies were state enterprises, the others being investor-owned.

The top 20:
Saudi Aramco, Chevron, Gazprom, ExxonMobil, National Iranian Oil Company, BP, Royal Dutch Shell, Coal India, Pemex, Petróleos de Venezuela, PetroChina, Peabody Energy, ConocoPhillips, Abu Dhabi National Oil Co, Kuwait Petroleum Corp, Iraq National Oil Company, Total S.A., Sonatrach, BHP Billiton, Petrobras.

Michael Mann, one of the world's leading climate scientists, said it was 'a great moral failing of our political system' for this to have been allowed to happen.

'The great tragedy of the climate crisis is that seven and a half billion people must pay the price – in the form of a degraded planet – so that a couple of dozen polluting interests can continue to make record profits.'

When reporting the investigation the *Guardian* newspaper approached all the twenty companies involved, and some of the seven who replied argued that they were not responsible for how the fuels they extracted were used. All claimed they were making efforts to invest in renewable or low carbon energy sources.

'Believe in the power of your own voice'

CHAPTER FOUR

SPEAKING OUT

On an August day in 2018 a small, pig-tailed 15-year old girl sat alone on the cold concrete outside the Swedish parliament in Stockholm with a handwritten placard announcing 'Skolstrejk för Klimatet', or 'School Strike for the Climate'.

With a speed that only social media can achieve, Greta Thunberg's principled truanting inspired young people around the world to follow suit – until, in September 2019, a little more than a year since that first lonely vigil, millions of them were marching with banners in 185 countries.

Thunberg, by now 16, sailed the 4,800 km (3,000 miles) from England's south coast to New York in a small emissions-free boat in order to make her case before Congress and the UN Climate Action Summit (*page 72*).

Blessed, as she believed, with Asperger's syndrome – a 'superpower', she said, that made her see things in black and white terms and 'from outside the box' – she spoke directly and without any semblance of fear to audiences of older people in high positions, many of whom she knew to be hostile to her views.

'I don't want you to listen to *me*,' she told US senators. 'I want you to listen to the scientists.'

And her anger spilled over after sitting through the summit and finding that the delegates had failed to rise to the challenge.

'This is all wrong,' she said, her voice cracking and tears coming to her eyes. 'I shouldn't be up here. I should be back in school on the other side of the ocean. Yet you all come to me for hope? How dare you! You have stolen my dreams, my childhood, with your empty words.

'And yet I'm one of the lucky ones. People are suffering. People are dying. Entire ecosystems are collapsing. We are in the beginning of a mass extinction. And all you can talk about is money and fairy tales of eternal economic growth. How dare you!

'If you choose to fail us, I say we will never forgive you.'

Legal redress

After the climate summit Greta Thunberg and fifteen other young activists announced that they were suing five countries they regarded as major carbon polluters – Argentina, Brazil, France, Germany and Turkey – with the aim of compelling them to work with other nations to forge binding emissions reduction targets.

Many US states already faced legal action by young campaigners, and in November 2018 the Supreme Court allowed a major lawsuit to go ahead against the US government – which had given notice of appeal on the grounds that there was no constitutional right to an environment free of climate change.

Sticks and stones

Inevitably Thunberg was belittled and condescended to. As she set off in her sailing boat for America Arron Banks, the British businessman and political donor, observed that freak yachting accidents weren't unknown, while Donald Trump, having witnessed her later tears, tweeted sarcastically 'She seems like a very happy young girl looking forward to a bright and wonderful future. So nice to see!'

The Fox News host Laura Ingraham suggested that she and her fellow young activists were 'brainwashed'; the pundit Michael Knowles described her on the network as 'a mentally ill Swedish child'; and the right-wing commentator Dinesh D'Souza compared her image to those used by Joseph Goebbels in his Nazi propaganda – 'notably Nordic white girls with braids and red cheeks'.

Thunberg's response was typically blunt: 'When haters go after your looks and differences,' she posted on Instagram, 'it means they have nowhere left to go. And then you know you're winning!'

Winning or not, the prophets of impending doom were becoming more vociferous and, to all but those determined not to be persuaded, more compelling.

Scientists had, over several decades, provided mounting evidence of man-made climate change, but the dry figures, the tables and the graphs had passed many people by. Now there were voices speaking out in an often powerful and eloquent language that couldn't be ignored.

A veep's cause

One of the earliest was the former US vice president Al Gore, who, having lost to George W. Bush in the race for the White House, dedicated his life to the environmental cause he had been championing for decades.

His book, *An Inconvenient Truth*, was published in 2006, and the next year he shared the Nobel Peace Prize with the IPCC 'for their efforts to build up and disseminate greater knowledge about man-made climate change, and to lay the foundations for the measures that are needed to counteract such change.'

A film with the same title included a slide show that Gore presented around the world more than a thousand times, and he followed this up in 2017 with *An Inconvenient Sequel: Truth to Power*, a film that showed him evangelising on his travels.

Many felt he was much more effective in his retirement than when, as vice president, he was compromised by the realities of US politics and watered down the terms of the 1997 Kyoto Agreement on greenhouse gases.

Silent Spring

A seminal natural sciences book of the present age was *Silent Spring*, written by the American biologist and conservationist Rachel Carson in 1962 and still in print today. It wasn't a climate change book (the world hadn't yet woken up to the problem), but tackled the damage caused to the countryside by the indiscriminate use of pesticides. The book, met with hostility by the chemicals industry, led to a US ban on DDT for agricultural purposes in 1972 (the UK followed suit in the 1980s) and inspired a new generation of environmentalists.

SPEAKING OUT

> 'Believe in the power of your own voice. The more noise you make, the more accountability you demand from your leaders, the more our world will change for the better.'
>
> *Al Gore*

The Green New Deal

The journalist Thomas Friedman, writing in the *New York Times* in 2007, was an early champion of a Green New Deal* – an echo of Franklin D. Roosevelt's ambitious New Deal of the 1930s, which helped lift the country out of the depths of the Great Depression.

'If you have put a windmill in your yard or some solar panels on your roof, bless your heart,' he wrote. 'But we will only green the world when we change the very nature of the electricity grid – moving it away from dirty coal or oil to clean coal and renewables.'

---

* *The idea was taken up in the UK, with the publication in 2008 of a report 'A Green New Deal: Joined-up policies to solve the triple crunch of the credit crisis, climate change and high oil prices.'*

Friedman emphasised the scale of what was needed: 'That is a huge industrial project – much bigger than anyone has ever told you. Finally, like the New Deal, if we undertake the green version, it has the potential to create a whole new clean power industry to spur our economy into the 21st century.'

Just how huge a project it would be emerged when the Green New Deal idea was taken up by a group of Democrats in 2018. The New York representative Alexandria Ocasio-Cortez (AOC for short) called for an electrical grid running on 100 per cent renewable energy and an end to fossil fuel use within ten years – an ambition costed at $2.5 trillion a year.

AOC argued that it should be financed by higher taxes, describing climate change as 'the single biggest national security threat for the United States and the single biggest threat to worldwide industrialised civilisation'.

She called for 'more environmental hardliners in Congress', but the first Green New Deal legislation was rejected by the Senate in March 2019.

A house on fire

'In a North American context,' the activist and author Naomi Klein told the media, 'it's the greatest taboo of all to actually admit that there are going to be limits.'

House style

In May 2019, aware that its references to climate change were often too bland, the *Guardian* newspaper amended its style guide. Although 'global warming' wasn't banned, journalists were advised that 'global heating' gave a better idea of the dangers involved.

As for 'climate change', the editor, Katharine Viner, declared that 'it sounds rather passive and gentle, when what scientists are talking about is a catastrophe for humanity'. The preferred terms were 'climate emergency', 'climate crisis' or 'climate breakdown'.

The paper also began to include among its weather forecasts the latest CO_2 ppm readings from Mauna Loa, comparing them with the pre-industrial base and the 'safe level' – left behind long ago.

10 books on climate change

The Uninhabitable Earth, David Wallace-Wells, Allen Lane

This Changes Everything: Capitalism vs. the Climate, Naomi Klein, Simon & Schuster

Six Degrees: Our Future on a Hotter Planet, Mark Lynas, Fourth Estate

Climate Change: What Everyone Needs to Know, Joseph Romm, OUP

Climate Change: A Very Short Introduction, Mark Maslin, OUP

The Great Derangement: Climate Change and the Unthinkable, Amitav Ghosh, University of Chicago Press

The Sixth Extinction: An Unnatural History, Elizabeth Kolbert, Henry Holt

The Weather Makers: How Man is Changing the Climate and What It Means for Life on Earth, Tim Flannery, Grove Press

Eaarth: Making a Life on a Tough New Planet, Bill McKibben, Henry Holt

An Inconvenient Truth: The Planetary Emergency of Global Warming and What We Can Do About It, Al Gore, Rodale Press

The country's history and its idea of itself made it difficult to argue against vital reforms which involved cuts in consumption.

'You see that in the way Fox News has gone after the Green New Deal – they're coming after your hamburgers! It cuts to the heart of the American dream – every generation gets more than the last, there is always a new frontier to expand to, the whole idea of settler colonial nations like ours.

'When somebody comes along and says, actually, there are limits, we've got some tough decisions, we need to figure out how to manage what's left, we've got to share equitably – it's a psychic attack.'

And in her 2019 book *On Fire: the Burning Case for a Green New Deal*, Klein admitted being puzzled: 'For a decade and a half, ever since reporting from New Orleans with water up to my waist after Hurricane Katrina, I have been trying to figure out what is interfering with humanity's basic survival instinct – why so many of us aren't acting as if our house is on fire when it so clearly is.'

Trumping climate action

The obstacles to checking climate change in countries with obstructive governments is exemplified by Donald Trump's presidency. Apart from withdrawing the US from the Paris Agreement, he sidelined scientists in various government departments.

• Qualified researchers were removed from boards reporting to the Environmental Protection Agency (EPA) and replaced by political appointees, including people from the fossil fuel industries and climate change deniers.

• A report on the effects of climate change on the country's 118 coastal parks was edited to remove any mention of man-made changes.

• The scientist hired to run the Office of Policy Analysis and whose speciality was climate change in the Arctic was moved from his role after speaking on the subject at the United Nations. He said the mantra of the administration was to 'drain the swamp' – 'and it quickly became clear that the swamp was not the high-paid lobbyists but the rank-and-file government professionals.'

- A laboratory engineering team working on greenhouse gas standards for vehicles had produced thousands of pages of analysis when the administration not only proposed an eight-year freeze of standards but 'cooked the books and changed every assumption to get the answer they wanted'.

- A water quality official at the EPA had a chemical safety rule she was working on re-written by someone from the industry lobby group, the American Chemistry Council.

- A scientist modelling sea-level rise over the next hundred years at the EPA was asked to leave, commenting later, 'Any time you see a report come out of the agency that even mildly mentions climate change the administration tries to downplay the human connection and federal scientists' work.'

These cases were all revealed by whistleblowers, one of whom said, 'The Trump administration is threatened by expertise. It's a profound threat to democracy but it also increases risks to American health and safety. Every American should be concerned.'

> 'The scientific evidence is that if we have not taken dramatic action within the next decade, we could face irreversible damage to the natural world and the collapse of our societies.'
>
> *Sir David Attenborough, 2019*

Klein took comfort from the emergence of the New Green Deal: 'I feel a tremendous excitement and a sense of relief that we're finally talking about solutions on the scale of the crisis we face – that we're not talking about a little carbon tax, or a cap and trade scheme, as a silver bullet.'

Pointing the finger

In the UK, where successive governments had shown themselves to take climate change seriously, writer-activists such as George Monbiot see it as their role to keep them up to the mark.

'Tinkering at the edges of this problem got us into this mess,' he has said. 'It will not get us out.'

And although acknowledging the problem to be essentially political, Monbiot casts his criticism more widely.

'It is not just governments that have failed to respond, though they have failed spectacularly. Public sector broadcasters have deliberately and systematically shut down environmental coverage, while allowing the opaquely-funded lobbyists that masquerade as think tanks to shape public discourse and deny what we face.'

The referee's whistle

In 2018 the BBC, which has always prided itself on impartiality, announced a change in policy regarding climate change. Reacting to criticism, it warned its staff to be aware of 'false balance'.

'You do not need to include outright deniers of climate change in BBC coverage,' ruled Fran Unsworth, the director of news and current affairs, 'in the same way you would not have someone denying that Manchester United won 2-0 last Saturday. The referee has spoken.'

'Academics,' Monbiot added, 'afraid to upset their funders and colleagues, have bitten their lips.'

He made a short film, *Protect/Restore/Fund*,* with Greta Thunberg, the pair enforcing their argument though a series of bullet points:

- Trees were nature's 'magic machine', sucking CO_2 from the air, costing very little and building themselves.

- Fossil fuels must be left in the ground.

- Up to 200 species were being lost every day.

- Tropical forests were being destroyed at the rate of 30 football pitches a minute.

- A thousand times more was being spent on subsidising fossil fuels than on solutions to climate change.

---

* The director, Tom Mustill of Gripping Films, said 'We tried to make the film have the tiniest environmental impact possible. We took trains to Sweden to interview Greta, charged our hybrid car at George's house, used green energy to power the edit, and recycled archive footage rather than shooting new.'

SPEAKING OUT

National treasure

Where Monbiot was happily combative and controversial, Sir David Attenborough was a warmly regarded naturalist who in his younger days had travelled into the distant wilds to share the TV cameras with mountain gorillas and creatures of the deep.

There were a few critics who worried that his colourful celebrations of nature's wonders ignored their growing fragility, but as he approached his nineties he tackled the issues of the day – the scourge of plastics in the oceans and the unignorable perils of global warming and climate change.

Breathing easily

We each breathe in about 9.5 tonnes of air a year, but oxygen accounts for only about 740kg (1600lbs) of that. If we want to be self-sufficient we should grow seven or eight mature trees – the number required to produce enough oxygen to fill one person's lungs.

The *Daily Telegraph* praised his 2019 BBC film *Climate Change – the Facts* in terms which emphasised his role as a national treasure: 'At a time when public debate seems to be getting ever more hysterical, it's good to be presented with something you can trust. And we all trust Attenborough.'

Rumbles down under

But he wasn't simply a cuddly figure. In September 2019, describing the bleaching of the Great Barrier Reef as 'a tragic sight', he appeared on Australian television to attack the Scott Morrison government for opening new coal mines.

'You are the keepers of an extraordinary section of the surface of this planet, including the Barrier Reef,' he said, 'and what you say, what you do, really, really matters.

'And when you've been upstanding and talking about what I see as the truth and you so clearly say, "No, it doesn't matter... it doesn't matter how much coal we burn... we don't give a damn what it does to the rest of the world."'

Morrison, whose country's coal exports totalled US$47 billion a year,* and who had, before being elected, provocatively waved a lump of coal in parliament, saying 'This is coal – don't be afraid,' immediately hit back.

'I think there's a lot of disinformation out there about, frankly, what Australia is doing,' he said.

He was in the US where, having not been invited to speak at the climate action summit because of Australia's emissions record, he instead visited Chicago to tour a drive-thru McDonald's.

Morrison couldn't resist a swipe at Greta Thunberg and her young activist colleagues, warning them against 'needlessly' worrying about the climate.

'Let kids be kids,' he said. 'I think it is important to give them that confidence that they will not only have a wonderful country and pristine environment to live in, but that they will also have an economy to live in as well.'

* Making it easily the world's largest exporter of coal.

Through the fire

Back in 2001 the right-wing political consultant and climate change denier Frank Luntz advised President George W. Bush how to defeat calls for clean energy legislation.

'There is still a window of opportunity to challenge the science,' he wrote in a memo. 'You need to continue to make the lack of scientific certainty a primary issue in the debate.'

At 3.15 on a December morning in 2017 Luntz woke to an emergency: he and his neighbours had to evacuate their Los Angeles homes because they were in the path of the Skirball Fire, one of a series of wildfires that destroyed parts of California that month.

Luntz referred to that dramatic event when appearing before a House select committee on the climate crisis in 2019. By now he was a believer in the science he had previously attempted to besmirch.

'I'm here before you to say that I was wrong in 2001,' he told senators. 'That was a lifetime ago... I've changed.'

Meanwhile, COAL 21, a research organisation established to develop low-carbon emission fuel, announced plans to fund a multi-million-dollar media campaign to make Australians feel 'proud about coal'.

Holy war

Increasingly outspoken about the perils of climate change, Pope Francis, leader of the world's 1.2 billion Roman Catholics, called any failure to act urgently against the production of greenhouses gases 'a brutal act of injustice toward the poor and future generations'.

He was speaking at a Vatican meeting with some of the world's leading oil companies, where he urged them to hear 'the increasingly desperate cries of the Earth and its poor'.

How did his guests react? They called on governments to put carbon pricing in place to encourage low-carbon innovation, and they asked for greater financial transparency to aid investors – but they made no pledges about reducing CO_2 emissions and they set no timetable for action.

And BP's chief economist warned later – indicating the tough battles that lay ahead – that the world 'should not rely on the private sector' when investing in clean technologies. Instead, governments should come to the companies' aid with taxpayers' money.

The environmental activists Greenpeace UK (at that moment protesting about fossil fuel emissions by occupying a BP oil rig in the North Sea) were predictably unimpressed.

The Green Pope

It was Pope Francis's predecessor, Benedict XVI, whose climate change credentials earned him the nickname the Green Pope.

'If you want to cultivate peace,' he said, 'protect Creation'. With the aim of making the Vatican the world's first carbon neutral state (made easier, it's true, by its tiny size), he had solar panels installed on the roof of the audience hall and introduced electric cars to his transport fleet. He also accepted land in Hungary's Bükk National Park to offset the Holy See's fossil fuel usage, although it seems that this gift never materialised.

SPEAKING OUT

> 'Never have we so hurt and mistreated our common home as we have in the last two hundred years.'
>
> *Pope Francis*

'The oil majors knew all about the risk from climate change many years before most of us first heard about it,' claimed their spokesman, Mel Evans. 'They knew their products were the cause, and yet they kept it quiet and lobbied for business as usual.

'And they're still lobbying for business as usual. When it comes to saving the planet they will do what they are forced to do and no more.'

Trouble with Harry

Those who speak out against the polluters are always in danger of being called out for hypocrisy. After issuing his environmental encyclical, *Laudato si'*, in 2015 Pope Francis went on a whirlwind tour of Latin America which was undoubtedly carbon footprint heavy. Indeed, during that year he visited 15 countries in five continents.

Lisa Sideris, director of the Indiana University Consortium for the Study of Religion, Ethics and Society, put up a 'utilitarian' defence of the pontiff.

'The good that he does by raising awareness of climate change,' she said, 'particularly given the pope's great symbolic significance, outweighs the carbon expenditure his travels entail.'

Prince Harry, the Duke of Sussex, got off far less lightly after making a series of green comments in 2019 – and his critics brandished some damning evidence against him.

'There's an emergency,' he said while visiting Botswana to help create a new forest habitat after decades of deforestation. 'It's a race against time and one which we are losing. Everyone knows it. There's no excuse for not knowing that.

'I don't understand how anyone in this world, whoever you are – you, us, children, leaders, whoever it is – no one can deny science, otherwise we live in a very, very troubling world.'

SPEAKING OUT

What seemed to provoke most outrage was his revealing in a *Vogue* interview with the primatologist Jane Goodall that he and his American wife Meghan had decided to have no more than two children* because of concerns about the environment.

Not only was this thought to be 'preachy', but it conflicted with the incessant flights the couple and their baby son had been taking, including four private jet trips within 11 days.

The *Guardian*'s former environment editor, John Vidal, gave them a comparatively polite lecture in the paper:

'To make any difference to planet Earth, you and your family really must stop taking those private jets to Jamaica, the luxury safaris in Botswana, the weddings in Montego Bay, the impromptu winter gateways in Tromsø, the 'babymoons' in Australia and New York, the downtime on Mediterranean islands and the quick flights to Fiji.'

---

* *The world's population is expected to rise to just under 11 billion by 2100, putting a severe strain on resources.*

The worry scale

A YouGov poll published in September 2019 compared attitudes towards climate change across 28 countries and regions.

Percentage believing that human activity is mainly responsible:

India 71
Thailand, Spain and
 Indonesia 69
Italy 66
Vietnam 64
Philippines 62
Singapore 54
Taiwan 53
Qatar, Kuwait and
 UAE 52
Great Britain 51
Hong Kong 50
Finland and Germany 49
France 48
Bahrain 46
China 45
Australia 44
Oman 43
Egypt 42
Denmark 40
USA 38
Sweden 36
Saudi Arabia and
 Norway 35

Percentage believing it will have a big impact on their lives:

Philippines 75
Vietnam 74
India 70
Qatar 65
Egypt 58
UAE 56
Kuwait and Thailand 55
Bahrain 53
Malaysia 47
Oman 46
Indonesia 45
Singapore and Saudi
 Arabia 41
Taiwan 38
Spain 32
Italy and Australia 29
China and France 26
Hong Kong 25
USA 24
Great Britain 17
Germany 16
Finland 14
Norway 12
Sweden 11
Denmark 10

Percentage believing a drastic change is needed to prevent the worst effects:

Spain 82
Qatar 80
China 79
Italy 78
Taiwan 76
Singapore, Hong Kong, Oman and Vietnam 74
Indonesia 71
Malaysia and Egypt 68
Kuwait and Bahrain 67
Great Britain 66
Thailand 64
Philippines and UAE 63
India and Australia 61
France and Saudi Arabia 58
Germany, Denmark and Norway 57
Finland 56
Sweden 55
USA 50

Percentage believing their own country should be doing more to prevent climate change:

India 22
France 20
Thailand 18
UAE 16
Hong Kong 15
Denmark, Italy, Norway, Philippines, Saudi Arabia 14
Taiwan 13
Egypt 12
Finland, and **Great Britain 11**
Australia and USA 10
Kuwait and Vietnam 9
Indonesia, Singapore, Spain and Sweden 8
Qatar 7
China and Malaysia 6
Oman 4

Few of those interviewed believe that climate change wasn't happening at all, with most doubters being found in the US (6 per cent). Nine per cent of Americans who *did* believe in it thought human beings were not at all responsible, followed by Norway (8 per cent), Egypt and Saudi Arabia (7 per cent).

Other savagers of what they saw as the prince's hypocrisy were far less well mannered, forcing him to admit, 'We can all do better... No one is perfect.'

The spat produced some interesting facts and figures:

- **By producing one child fewer, parents reduce their CO_2 emissions by 58.6 tonnes a year – around 25 times more than from any other course of action, such as selling the car.**

- **Taking four journeys by private jets consumes 82 tonnes of CO_2, roughly equivalent to the average use of 17 cars for a year.**

Responsibilities

The Harry/Meghan controversy takes us to the very nub of the climate change crisis. As royals with money to burn, their 'norms' are of course different from ours. But if we contrast their obvious good intentions with their unthinking acceptance of a comfortable status quo, don't we find it echoed in our own reluctance to change the habits of a lifetime?

SPEAKING OUT

If the Earth is to be saved from a sixth mass extinction and if a malign wretchedness is not to be visited upon the human race across the globe within the lifetime of our youngest, vast changes have to be made to the way we live.

How urgently are we prepared to make them?

The nay-sayers

Although a welling tide of voices has been raised in warning of future climate change meltdown, there are still those who deny human responsibility for the crisis – and some of them wield significant power and influence.

President Trump, who has described climate change as a 'hoax', responded to a government report that found unchecked global warming would wreak havoc on the US economy by saying simply, 'I don't believe it'. His vice president, **Mike Pence**, painted the best gloss he could on the official line ('Our president is choosing to put American energy and American industry first'), while the White House chief of staff, **Reince Priebus**, attempted to put the dangers of climate change into perspective: 'Look, I think we all care about our planet, but melting icebergs aren't beheading Christians in the Middle East.'

The Russian leader, **Vladimir Putin**, at one time acknowledged a 'climate problem', but in 2017 he alluded to the observations of a 1930s Austrian explorer with a photographic memory in order to claim that icebergs had been melting for decades from natural causes.

He welcomed global warming for exposing transport routes and natural resources which had hitherto been too expensive to exploit, and suggested that the rise in temperatures would benefit Russians who would no longer need to wear fur coats.

With scientists reporting that 15 per cent of the Amazonian rain forest had already been lost, and that by 2021 it was likely to reach the 'tipping point' after which it would be unable to generate sufficient rain to sustain itself, Brazil's president, **Jair Bolsonaro**, claimed that the forest was 'practically untouched' and blamed 'a lying and sensationalist media' for spreading fake news. His foreign minister, **Ernesto Araújo**, denied that there was a climate change catastrophe, adding that 'from the debate that is going on it would seem that the world is ending'.

In the UK the former chancellor of the exchequer Nigel Lawson, now **Baron Lawson of Blaby**, founded the Global Warming Policy Foundation, a think tank sceptical of climate change science. 'The next government,' he said in 2017, 'must take a long, hard look at whether we can afford our own Climate Change Act any longer. It is clear that the costs imposed on British businesses and households are now entirely unsustainable.'

"Future generations will almost certainly look back and wonder why on earth – why on Earth – did we choose our suicide?"

CHAPTER FIVE

DESPERATE MEASURES

Economists have laboured long and hard to work out the global costs of combating climate change (and their estimates have veered so wildly as to be practically useless), but the very question seems pointless. Can there be any cost too great if the alternative is, in the words of the International Monetary Fund's former chairman, Christine Lagarde, being 'roasted, toasted, fried and grilled'?

We already have much of the necessary technology, but what impedes the carbon-free revolution is a fear of the collateral damage it will inflict – on our pockets and our lifestyles.

The likes of us

As individuals we can make our own small difference. A vegan hermit would smile at the list below, but for many of us some of these restrictions would prove highly inconvenient even if we could afford them.

- Don't fly.

- Walk and take public transport rather than use the car – which should be electric.

- Install solar panels on your roof.

- Use low-energy light bulbs.

- Plant some trees.

- In cold weather, wear more clothes indoors rather than turn on the central heating.

- Dry washing outside on an airer rather than using a tumble drier.

- Recycle more.

- Eat locally sourced food.

- Cut meat and dairy products from your diet.

DESPERATE MEASURES

> '*Future generations will almost certainly look back and wonder why on earth – why on Earth – did we choose our suicide?*'
>
> *Jonathan Safran Foer*

Those addicted to a meat diet may like a tip from Jonathan Safran Foer, the American novelist and a vegan keen to stress that 'animal agriculture produces more greenhouse gas emissions than the entire transportation sector (all planes, cars and trains), and is the primary source of methane and nitrous oxide emissions' – as well as causing mass deforestation.

Arguing that the average US and UK citizen should consume 90 per cent less beef and 60 per cent less dairy, Foer has a simple compromise suggestion: eat no animal products for breakfast or lunch.

'I bet,' he has written, 'that if most people think back over their favorite meals of the past few years – the meals that brought them the most culinary and social pleasure, that meant the most culturally or religiously – virtually all of them would be dinners.'

Losing the crap

The American engineer Saul Griffith was so concerned with the practical challenge of lowering his carbon footprint that he worked it out in amazing, nerdy detail. He not only recorded how much energy he used in travelling and in running equipment at home, but how much energy was needed to transport everything he used – including a catamaran he had to replace every two years.

While the 'energy budget' of the average person in the world was about 2,200 watts, he said, he was consuming 18,000 watts, 'like most Americans'.

He decided that 'a quarter of the energy we use is just in our crap' and drastically reduced his own output by cutting most of his air travel; using a bike rather than the car whenever possible; keeping to the speed limit when driving (he hadn't overtaken anyone in months); eating meat once a week; and buying almost nothing.

And yes, he said it had made him healthier and happier!

However hard we try to reduce our personal carbon footprint, however, it's a hard truth that this is merely tinkering with the problem. The coming catastrophe can only be averted if governments around the world take much more drastic action than any of the major players have so far shown themselves willing to do.

It seems that the short-term financial and political pain outweighs the real but relatively distant disaster.

Eeyore's warning

In June 2019, in the dying days of Theresa May's premiership, her gloomily visaged chancellor of the Exchequer, Philip Hammond (commonly known as Eeyore after Winnie the Pooh's friend), gave her a glum report on climate change progress.

Although the independent Committee on Climate Change (CCC) was arguing for zero-carbon emissions by 2050, the government was still working towards a figure of 80 per cent – and Hammond admitted that it had fallen behind even that target.

Backing the polluters

Despite the UK's commitment to the Paris agreement, a report from the European Commission in 2019 revealed that the country gave more subsidies to fossil fuels than any other member of the EU.

While it backed renewable energy to the tune of about £7.5bn a year, its support for fossil fuels amounted to more than £10bn.

VAT on domestic electricity and gas was fixed at 5 per cent rather than the prevailing 20 per cent to help homeowners, while motorists had benefited from a freeze on petrol and diesel duty which the chancellor of the exchequer, Philip Hammond, admitted had cost some £46bn in lost revenues between 2011 and 2019 – 'about twice as much as we spend on all NHS nurses and doctors each year'.

The government argued (to widespread condemnation) that these measures didn't constitute fossil fuel subsidies according to its own definition of the term, but the Commission's report warned that 'EU and national policies might need to be reinforced to phase out such subsidies'.

DESPERATE MEASURES

The CCC reckoned it would cost £50 billion a year to be carbon free by 2050, but Hammond believed an annual figure of £70 billion was more realistic – with a total bill 'well in excess of £1 trillion' – and some industries would find themselves 'economically uncompetitive' unless other countries followed suit.

Regarding it as pure cost, rather than an investment with some financial benefits, he told the prime minister it would mean less money for schools, hospitals and the police, and that the implications of pursuing the target needed to be 'better understood'.

His dogged caution underlined the difficulty of taking hard decisions in a democracy.

After all, those vital public services could be maintained *and* the government could meet its carbon free targets if the voting public was prepared for job losses and higher taxes – if, that is, today's generation was willing to take a hit on behalf of the health and security of generations yet to come.

Would you sign up for that?

Healing the ozone

The story of the hole in the ozone layer (caused by human technology, risking the lives of millions and involving a long battle between scientists and industrialists who disputed their findings) is a forerunner of the climate change crisis – and, thanks to international action, it has a happy ending.

In 1974 the chemists Sherwood Rowland and Mario Molina discovered that the CFCs (chlorofluorocarbons) in aerosol sprays were rising high into the stratosphere where, through a chemical reaction, the Sun's rays destroyed the vital shield of ozone molecules protecting us from ultra violet (UV) rays that can cause skin cancers and blindness from cataracts.

The public rapidly turned to pump sprays and roll-ons, and, although some manufacturers denied the science – 'A science fiction tale... a load of rubbish... utter nonsense,' exploded the chair of DuPont – measures protecting the ozone layer were added to the US Clean Air Act. In 1980 it was found that other uses of CFCs in refrigerators, air conditioners and industrial processes were also a threat to public health.

The battle wasn't over, however. Action stalled during Ronald Reagan's first term as president, and it took a concerted effort by congressmen from both major parties to create a sense of urgency – helped in the mid-1980s by the frightening appearance of a hole in the ozone layer over the Antarctic.

A plan was agreed to phase out the CFCs over a ten-year period, although some die-hards continued to fight against it. Reagan's interior secretary Donald Hodel (a figure akin to today's industry-backed climate change deniers) urged the president to tell people to protect themselves with hats, sun lotion and sunglasses. Reagan happened to suffer from skin cancer himself, and he knew where the truth lay.

- In September 1987 countries around the world agreed on the Montreal Protocol, and it was most recently strengthened by the Kigali Amendment in 2016.

- In 1995 Sherwood Rowland and Mario Molina were awarded the Nobel Prize in Chemistry.

- The ozone layer is healing and the Antarctic hole is expected to close up later this century.

Keeping it in the ground

The one single measure that would give those future generations a fighting chance is simple to understand and yet probably the most difficult to achieve politically – leaving all the untapped fossil fuels where they are.

Back in 2014, before the Paris summit, the governor of the Bank of England, Mark Carney, went so far as to warn companies and investors that 'the vast majority of reserves are unburnable'. It's been called the 'carbon bubble theory'.

Such was the risk of disaster, he said, that businesses should be aware of what he called their 'stranded assets'. He encouraged more of them to include environmental inputs and investments alongside their annual reports to shareholders. *That seems a long time ago.*

Bill McKibben, the American environmental activist and author, has been a prime mover in the Keep It In The Ground protest movement – and he's under no illusions about the scale of the challenge he and his colleagues face.

'There's oil in the Arctic,' he's written, 'and in the tar sands of Canada and Venezuela and in the Caspian Sea; there's coal in Western Australia, Indonesia, China and in the Powder River Basin [in Montana and Wyoming]; there's gas to be fracked in Eastern Europe.

'Call these the "carbon bombs". If they go off – if they're dug up and burnt – they'll wreck the planet. Of course you could also call them "money pits"... That coal and gas and oil may be worth $20tn, maybe more.

'Because of that, there are people who say that the task is simply impossible, that there's no way the oil barons and coal kings will leave those sums underground. And they surely won't do it voluntarily!'

Sticks and carrots

While many companies have adopted green credentials, governments have for years run schemes aimed at reducing carbon emissions through persuasion rather than edict – although the side effects haven't always been in the best interests of the planet.

Under so-called **Cap and trade** – a form of 'carbon trading' – companies are given a limit for the amount of carbon they are allowed to produce. If they can't manage it they can buy 'carbon credits' from less polluting companies.

Getting the shakes

Fracking has made the US the world's leading producer of shale gas, but in the UK limited trials have produced angry protests.

Huge volumes of water, sand and what critics regard as potentially dangerous chemicals are pumped underground to displace the gas. A side effect can be earth tremors, and the company involved, Cuadrilla, closed down its 2019 operation after recording three of them in less than a week – one registering 2.9 on the Richter scale.

The Conservative government not only defended its pro-fracking policy against charges that it was sending yet more carbon into the atmosphere, but introduced new, relaxed planning measures which, in the words of Greenpeace, would make exploratory drilling 'as easy as building a garden wall or a conservatory'.

DESPERATE MEASURES

What the critics say: The dirtiest firms simply buy the credits they need and carry on just as they did before.

Carbon offsetting is a system that enables people or companies to 'neutralise' their carbon emissions.

If you're taking a long-haul flight you can offset the damage caused to the environment by paying a fee that's used to invest in green projects – renewable energy, say, or the planting of trees.

What the critics say: It's a way of allowing the wealthy to carry on polluting guilt-free, and some of the green initiatives might have happened anyway.

Green taxes (or eco-taxes) raise money for beneficial causes or to steer people's behaviour in a more environmentally friendly direction. With a fuel tax, for instance, the more carbon you emit, the more you pay.

What the critics say: They hit the poor most, as they have less disposable income, and the cynical public often regard them simply as a convenient cash cow.

Money talks

A powerful new weapon against the oil, gas and coal producers began to be brandished on US university campuses in 2010 – fossil fuel 'divestment'.

Students urged their ruling bodies to convert their existing endowment holdings in the fossil fuel industry into investments in clean energy instead.

By September 2019 more than a thousand institutions and 58,000 individuals between them representing $11 trillion in assets throughout the world had been divested from fossil fuels.

The effect of bringing moral, political and economic pressure to bear on polluters is nowhere better seen than in the story of the long delayed and much reduced Carmichael mine in central Queensland, Australia.

In 2010 the Indian firm Adani announced plans to breach the massive untapped coal reserves of the region by developing the world's biggest mine, employing 10,000 people.

The protesters found many reasons to object – damage to the Great Barrier Reef; to the local environment; and to the cultural heritage of the land's traditional owners, the Wangan and Jagalingou People – but they also targeted the industry's weak spot.

Such vast projects as this one require huge injections of capital, and public pressure was put on Australia's big four banks not to put up the cash: they caved in. The federal government now considered granting a billion dollar concessional loan to get the scheme off the ground, but Queensland's premier bowed to public pressure and vetoed it.

Out, damned spot!

In October 2019, after months of campaigning by artists, environmentalists and the public, the Royal Shakespeare Company announced that it was ending its sponsorship deal with BP. 'Young people are now saying clearly to us that the BP sponsorship is putting a barrier between them and their wish to engage with the RSC,' said its artistic director, Gregory Doran. 'We cannot ignore that message.'

The environmental group Market Forces used its website to alert its followers to all the companies who were giving their support to the development, with a suggested letter they might send to each one of them:

> Re: Adani Carmichael coal project
>
> Given the services your company offers, I am concerned that you are working to support Adani's plans to mine and export coal from the Galilee Basin in Queensland. I want you to confirm to me personally, and announce publicly, that you will not be involved in any part of this project.

Adani wasn't beaten but it was badly wounded. In November 2018, having proved unable to raise the investments it needed, it settled on a self-funded and slimmed-down Carmichael project.

The Australian mining giant BHP, meanwhile, announced that it was turning away from coal in its portfolio because 'the world will be a very different place in 10 to 20 years' time'. Coal, it said, was likely to be phased out 'potentially sooner than expected'.

It was, indeed, already possible to envisage a world in which none of the fossil fuels played a major role – a world in which our energy needs were largely met by 'renewable' sources such as sunlight, wind and rain.

The Sun has got his hat on...

...as the song has it – and when he comes out to play he beams immense amounts of energy onto the land and sea. Here's how Mike Berners-Lee puts it in his book *There is No Planet B*. Solar energy is powerful enough:

- **For each of us to bring an Olympic swimming pool to the boil every day.**

- **For everyone on the planet to have 2,700 kettles on the go all the time.**

- **For the global population to be permanently in flight above the Earth ten times over.**

Harnessing this power is the challenge, and here some countries have a natural advantage: between them, Australia, Russia, China, Brazil and the US receive 36 per cent of all the sunlight that hits the Earth.

Second wind

There was a vogue for electric cars with rechargeable batteries in the late 19th and early 20th centuries, with as many as 30,000 of them on the roads in Europe and the US - and in 1899 the Belgian racing driver Camille Jenatzy broke the 100kph (62mph) speed barrier near Paris in his torpedo-shaped La Jamais Contente (or The Never Satisfied).

Mass produced petrol powered vehicles with electric starters soon proved more efficient, and the idea of electric cars ever making a comeback seemed a pipe dream until the early 21st century when pollution in cities and fears for the climate spurred investment in new technology.

By 2020 there were already more than 5 million electric cars on the road worldwide - proof that swift change is possible if led by government subsidies and public demand.

The UK's original target was to phase out petrol and diesel cars by 2040, although there were hints that this might be brought forward to 2035 in accordance with a recommendation of the independent Committee on Climate Change.

DESPERATE MEASURES

Those at the other, gloomier end of the scale include Bangladesh, Rwanda, South Korea, Japan, Belgium, the Netherlands and – you guessed it – the UK.

Engineers are beginning to meet the challenge with more efficient, and cheaper, batteries to store that captured energy. That's **Solar PV**, as in photovoltaic cells which directly convert sunlight into electricity. These cells are found in the panels on house and factory roofs* (with any excess electricity sold to the national grid), but larger versions are installed in expansive PV 'solar parks' in the world's hotspots.

Solar thermal systems, on the other hand, use the sunlight to heat liquids – grandly, as in commercial power plants (which create steam to run turbo generators), more humbly for direct cooking and heating in homes, many of them in the developing world.

---

* *In 2010 the UK government introduced a support scheme to encourage the installation of solar panels by homeowners, but these incentives were slashed severely in 2016 and cut altogether in 2019, with the result that installations plummeted and companies closed down.*

While vast solar plants and parks are being built in desert areas across the globe, the stresses involved in siting solar farms in the small-scale and shared English countryside are exemplified by plans to create one with 800,000 panels at Graveney in Kent – huge by UK standards. In late 2018 the government agreed to consider the proposal.

Bigger and bigger

Switched on in June 2019, the Noor Abu Dhabi solar plant became the largest anywhere in the world – although it didn't expect to hold the record for very long.

A consortium including companies from China, Japan and Abu Dhabi itself invested $870 million in the project, which has a total capacity of 1,177 MW and sells electricity to the Emirates Water and Electricity Company.

Meanwhile the Tengger desert solar park in the Ningxia region of China held the record as the world's largest PV plant. Its photovoltaic cells cover all of 43sq km (16sq miles) – and of course the array is known as the Great Wall of Solar.

The developers of the Cleve Hill Solar Park claimed that it would bring clean electricity to 91,000 homes, but there were protests on the grounds of ugliness ('an industrial landscape'; 'like a massive warehouse covering 600 football pitches') and its effects on wildlife, including overwintering birds.

favourable winds

What the UK lacks in sunshine it makes up for in wind – particularly offshore. As of 2019 it had 44 per cent of all European offshore wind power capacity, with two thousand turbines spread over 37 separate sites and providing 8GW of capacity.

As with solar, wind power projects have been growing both in number and in size. Back in 2013 the London Array in the outer Thames Estuary led the way, with 175 individual turbines and a capacity of 630MW. Within five years the Walney Extension in the Irish Sea had surpassed it, covering the equivalent of 20,000 football pitches, generating 659MW and supplying clean and cheap energy to 590,000 homes.

Destined to be the largest offshore wind farm in the world when it opens in 2023,* the Dogger Bank cluster in the North Sea is to have megablades potentially creating 5 per cent of the UK's total power supply.

At its 2019 conference, the opposition Labour party set a zero emissions target of 2030 for the UK should it come to power, and it followed this commitment with plans to build a further 37 offshore wind farms, so creating 700 jobs.

Water power

Although solar and wind lead the way, there's a range of other renewable sources – some of them, such as **Wave and tidal energy**, limited to specific locations. Typically a river estuary is dammed, and the incoming and outgoing tides drive turbines as they flow through gaps in the barrage. Another option (and less of a hindrance to ships and leisure boats) is to place turbines directly into the sea.

---

* *The Chinese, investing heavily in renewables, have the world's largest onshore wind farm at Gansu, with an eventual capacity of 20GW.*

farewell coal...

In June 2019 the UK's electricity grid operated for 18 days and 6 hours without coal – a record.

The percentages of energy used were: gas (40), nuclear (20), wind (13), imports (11), biomass (8), solar (7), hydro/storage projects (less than 1 per cent each).

The last of the country's coal-fired power stations were due to be phased out by 2025.

...but hello gas

After Drax Power applied to replace its existing coal-fired units in Yorkshire with four gas turbines (CCGTs) after 2025, an environmental assessment found that the project's scale, high emissions intensity and long operating life made it a 'significant threat' to the country's carbon targets.

The UK's Planning Inspectorate backed this report, but in October 2019 the Conservative secretary of state for business, Andrea Leadsom, overturned the decision and gave Drax the go-ahead.

The UK government backed research which helped make the country a world leader in ingenious technology – in twelve months a single floating turbine off the coast of Orkney generated more energy than the entire Scottish wave and tidal sector had managed in 12 years – but it then balked at the cost of the installations.

Shining lights

In the 1870s Joseph Swan and Thomas Edison developed the incandescent light bulb – and very thankful we've been to them ever since. But three less familiar names entered the pantheon in 2014, when Isamu Akasaki, Hiroshi Amano and Shuji Nakamura won the Nobel Prize in Physics.

Their achievement was devising the 'light-emitting diode' – or LED – light bulbs which, thanks to the measures taken by governments throughout the world, have rapidly been replacing the older technology.

Lighting, amazingly, accounts for well over a quarter of global electricity consumption, and the LED bulbs are both longer lasting and much more energy efficient.

DESPERATE MEASURES

A planned tidal lagoon power plant in Swansea Bay would have become the first of its kind in the world, operating for 14 hours a day and providing the power for more than 150,000 homes. It would also have been constructed to function as a protection against storms and floods.

The projected cost was £1.3 billion, and a killer blow was a (contested) report by the National Infrastructure Commission: 'Offshore wind becomes economic in 2030 without subsidies. Tidal never becomes competitive without government support.'

In 2018 the government withdrew that support.

Hydroelectricity – using running water to spin turbines downstream – is the leading source of renewable energy globally, with large-scale projects in China, Brazil, Canada, the US and Russia.

In the UK, where suitable sites are limited, it accounts for some 18 per cent of the country's renewable energy and less than 2 per cent of its energy supply overall.

If UK governments remain reluctant to invest in major schemes, it's suggested that much smaller-scale projects* could involve the repurposing of existing watermills, adding up to 2 per cent to the country's hydropower generation.

The nuclear option

Without a doubt the most controversial source of low-carbon energy is nuclear power, and this list is one reason why:

- Windscale 1957
- Three Mile Island 1979
- Chernobyl 1986
- Fukushima 2011

Yes, accidents can happen with nuclear fission (these are just the worst) – and when they do they can be very nasty indeed.

---

* *Hydroelectricity was first harnessed in Britain in 1878 to power the lights of the Victorian country pile Cragside in Northumberland. The home of the 1st Baron Armstrong, founder of the Armstrong Whitworth armaments firm, it's now owned by the National Trust.*

Bioenergy

Running cars on fuel made from food crops (or biomass) rather than petrol sounds a green idea, and it's popular in Brazil and the US, but there are snags. For one thing, there are disputes as to how much they cut down on greenhouse gases, and for another there's a concern about the acres they occupy. With farmland increasingly at a premium as climate change brings flooding and desertification, the spread of bioenergy probably isn't a very good idea.

Geothermal energy

We're familiar with the pumps which take natural heat from underground to warm houses and factories, but in some parts of the world large-scale plants use hot water or steam from below the surface in order to spin turbines and produce electricity. They provide a quarter of Iceland's total electricity demand, with figures of 22 per cent for El Salvador, 17 per cent for Kenya and the Philippines and 13 per cent for Costa Rica. The IPCC has forecast that geothermal could contribute 4 per cent to global energy use – a small, but useful, contribution.

There's a second knotty problem: what do you do with the highly radioactive waste that remains hazardous for tens of thousands of years? (The decommissioning facility at Sellafield* in Cumbria has been described as 'the most hazardous industrial building in western Europe'.)

On top of that, the plants use massive amounts of water (about to become an ever more precious commodity) and they're expensive.

As of 2020, 10 per cent of the world's electricity was generated by 450 nuclear power reactors in around 30 countries, with another fifty under construction, but after the Fukushima disaster in Japan several countries vowed never to go nuclear or to phase out the plants they already ran.

The UK, which launched the world's first civil nuclear programme at Windscale in 1956, expects to derive a third of its electricity from nuclear generation by 2035.

Formerly Windscale, but renamed in 1981.

Miniature Suns

A long nurtured dream among scientists has been mimicking what the Sun does – merging hydrogen atoms through nuclear fusion in order to release staggering amounts of energy with no nasty side effects.

A major problem has been achieving 'energy gain' (getting more out than the immense amounts you put in), but today this seems less of a wild idea. Several 'start-up' companies forecast that fusion could soon become a reality, and the British government has backed the development of a group of fusion reactors ('miniature suns') at the Culham Science Centre near Oxford.

'The basic science is solved,' said Wade Allison, emeritus professor of physics at Keble College, Oxford. 'The problems are technical ones to do with materials.'

The challenge, in short, is how to build reactors strong enough to contain the 'nuclear soup', or plasma, at scarcely imaginable temperatures and under immense pressure.

Thirty-five countries (including the EU) are involved in building the experimental International Thermonuclear Experimental Reactor (ITER) tokamak complex in Provence, southern France, but the signs are that private companies will beat them to it – and with much smaller, transportable devices.

A solution being worked on by Tokamak Energy in Oxfordshire involves using high temperature superconductors (HTS) to contain the plasma in a strong magnetic field.

'We expect to have energy gain capability by 2022,' said its CEO, Jonathan Carling, 'and be supplying energy to the grid by 2030.'

The company's aim was to hit temperatures of 180 million°F (100 million°C) – a target Chinese scientists claimed already to have achieved.

In the US, meanwhile, backed by billionaires such as Bill Gates, Jeff Bezos and Michael Bloomberg, a Massachusetts Institute of Technology (MIT) team is working on fusion reactors small enough to be built in factories and shipped for assembly on site.

DESPERATE MEASURES

A stitch in time

While we await the happy day when these renewable alternatives combine to replace fossil fuels, enterprising engineers, scientists and social scientists at Cambridge University are hoping to mend the damage we've already caused.

In 2019 they launched the Centre for Climate Repair (the first initiative of its kind in the world) by giving an idea of a few of the 'geoengineering' ideas they'd be looking at.

Refreezing the poles
The brainwave here is to brighten the clouds above them. Seawater would be pumped up through very fine nozzles fitted to tall masts on uncrewed ships. Tiny particles of salt would be injected into the clouds, making them more widespread and reflective – so cooling the areas below.

Greening the oceans
Fertilising the sea with iron salts could promote the growth of plankton that take up CO_2 and convert it into oxygen.

Previous experiments haven't worked, but Professor Callum Roberts told the media that approaches once thought beyond the pale now had to be given serious consideration.

'Early in my career,' he said, 'people threw their hands up in horror at suggestions of more interventionist solutions to fix coral reefs. Now they are looking in desperation at an ecosystem that will be gone at the end of the century, and now all options are on the table.'

Carbon capture and storage (CCS)

Sucking carbon out of the air has long been a pipe dream, and it may remain so, but at the launch of the Centre for Climate Repair Professor Peter Styring highlighted a pilot scheme at the Tata Steel plant in Port Talbot, South Wales, which recycles CO_2 before it escapes into the atmosphere.

'We have a source of hydrogen, we have a source of carbon dioxide, we have a source of heat and we have a source of renewable electricity from the plant,' he explained. 'We're going to harness all those and we're going to make synthetic fuels.'

Under arrest

CCS isn't a new idea (a pilot plant opened at the Vattenfall Schwarze Pumpe power station in Spremberg, north Germany, in 2008), but previous schemes have required a lot of energy and have been prohibitively expensive.

In the version illustrated below the captured carbon isn't turned into fuel but is pumped into underground oil and gas fields, where it's trapped by the rock above it.

How carbon capture and storage works

1. Mining of fuel
2. Coal- or gas-fired power station with CO_2 capture plant
3. CO_2 transport by pipeline
4. CO_2 injection
5. CO_2 storage sites

gas field

Depleted oil and gas fields

The Centre's coordinator, Sir David King, warned that time was no longer on our side.

'What we continue to do,' he said, 'what we do that is new, and what we plan to do over the next ten to twelve years, will determine the future of humanity for the next 10,000.'

A headline in the *Ecologist* journal over an otherwise positive report on the launch was telling: 'Welcome to the Centre for Climate Despair.'

from little acorns

Nature, as we've seen, already has its own carbon capture system in place: photosynthesis. In 2019 a scientific study used high-resolution satellite images from Google Earth and data covering soil, topography and climate factors in order to gauge how much land could be covered with trees without impinging on food production or living areas.

Its conclusion was that planting billions of trees across the world was the biggest and cheapest way of tackling the climate crisis.

DESPERATE MEASURES

Some 1.2tn native tree saplings would be planted on 1.7bn hectares (4.2bn acres) of bare land – an area equivalent to the size of the US and China combined. The cost was likely to be around $300bn (£240bn), perhaps met by government subsidies to landowners.

'What blows my mind is the scale,' said Tom Crowther, the ETH Zürich university professor who led the research. 'I thought restoration would be in the top 10, but it is overwhelmingly more powerful than all of the other climate change solutions proposed.'

It would, he said, take up to a hundred years for the scheme to have its full effect of removing 200bn tonnes of carbon from the atmosphere.

Although 48 nations had backed the so-called Bonn Challenge of 2011, aimed at restoring 350m hectares (865m acres) of deforested and degraded land by 2030, he thought this wasn't sufficiently ambitious because many of them had twice as much land available for fresh plantings: 'This is a new opportunity for those countries to get it right.'

Among the report's findings:

- In Europe the potential for new forests that avoid encroaching on cropland is highest in the UK, Ireland and central Europe.

- In the US, the temperate forest region of the south and east has the greatest potential.

- In Australia the tropical east coast is the most suitable.

Africa's Green Wall

Several countries have made substantial individual commitments to reforestation (China has introduced more than 50 billion trees since the 1970s, Australia intends to plant a billion by 2030 and Ireland 22 million each year over a 20-year period), but arguably the most dramatic response to land degradation and drought is the Great Green Wall on the arid savanna fringing the Sahara Desert.

Supported by more than twenty African countries, this $8 billion (£6.3 billion) scheme covers all of 100m hectares (247m acres).

Its aim is nothing less than transforming the lives of millions of people by sprouting a mosaic of green and productive landscapes stretching 15km (9 miles) from north to south and 8,000km (5,000 miles) from east to west – as much of it was back in the 1970s before population growth, poor farming and climate change denuded it.

And the camel came too

Teddy Goldsmith (1928–2009), the founding editor and publisher of *The Ecologist*, was ahead of his time in warning of pending environmental disaster.

Standing at the 1974 UK general election in Suffolk as a representative of the embryonic People Party,* he led a Bactrian camel on his campaign trail and had bearded supporters dress as Arab sheiks – the implication being that the county faced desertification thanks to oil-intensive farming practices.

Goldsmith lost his deposit, but his message doesn't seem as cranky today as it did then.

* It later became the Green Party.

Further objectives are the creation of 350,000 jobs and absorbing 250m tons of CO_2.

The project is only about 15 per cent complete, but drought-resistant acacia trees have been planted across the entire width of the continent, including an area of 75m hectares (30m acres) in Senegal alone.

In addition, many groundwater wells have been replenished with drinking water and food supplies have improved for rural towns.

Once finished, its champions claim, the Great Green Wall will be three times the size of the Great Barrier Reef and the largest living structure on the planet.

The gist of the matter

Imaginative, ingenious and invaluable as many of the initiatives outlined in this chapter undoubtedly are, they come with a heavy word of caution: none of them is sufficient by itself, and even in combination they won't be enough unless the essential problem of climate change is tackled at its root simultaneously.

Here's the conclusion of the 2019 annual report from the international energy consultants DNV GL: 'It is important to differentiate between the progress we are making on technology and the progress we are making on regulation.'

Its CEO, Ditlev Engel, told the media that 'Right now we are on a road to a place nobody wants to go' – a world, he predicted, heading towards a 2.4°C temperature rise by 2100.

'Business as *unusual*,' he said, 'has to become the new "business as usual". The investment houses I speak with are utterly confident about the technology; what they are concerned about is the regulation and where that is heading.'

The means were there, in other words – but where was the will?

Where your money goes

Despite a growing understanding of the perils of climate change, a 2019 report on investments made by money management firms revealed that they were pumping large sums from people's private savings and pension contributions into fossil fuel investment portfolios.

At a time when there were widespread calls for divestment,* the so-called 'Big Three' fund managers (BlackRock, Vanguard and State Street, whose combined assets outstripped China's GDP) had between them built a portfolio of public oil, gas and coal companies worth $300bn. That represented an increase of 34.8 per cent since 2016.

The *Observer* newspaper worked with the think tank InfluenceMap and the business data specialist Proxy Insight to carry out the investigation.

---

* *In June 2019 Legal & General, at one time a top 20 investor in ExxonMobil, announced that it was selling a $300m stake in the company and would use its remaining shares to vote against the chief executive.*

'While asset managers do not own the companies in which they invest,' it reported, 'they often exercise shareholders' rights on behalf of clients to vote on board members and company policy issues.

'Disclosures for publicly available company reports show that from 2015 to 2019 Vanguard and BlackRock used their votes to frequently oppose efforts to hold fossil fuel executives to account.'

None of the Big Three contested the findings, although they defended their policies:

- Vanguard said it was concerned about the long-term impact of climate risk and regularly engaged with companies on its shareholders' behalf.

- BlackRock said that it offered investors a wide range of environmentally sustainable investment options and was a leading investor in renewable power generation globally.

- State Street said that if an investor should want a strategy that considered climate issues 'we can provide that too'.

"We are the first generation to feel the sting of climate change, and we are the last generation that can do something about it."

CHAPTER SIX

THE DOOMSDAY DECADE

We are living through the Doomsday Decade. The world won't appear vastly different at the end of it, but what we do or don't do for the climate in 2020–2030 – as individual consumers, as campaigners, as industries, as governments – will affect the future of the globe and those who inhabit it for ever.

Fine words have been spoken and ignored, targets have been set and missed, and with every lost opportunity the chances of averting catastrophe have slipped further beyond our reach. These ten years are our testing time.

California greening

The persistent droughts and wildfires in California have driven the state to introduce a wide range of laws to minimise the effects of global warming and climate change, one of the most significant being a mandate requiring all new homes to incorporate solar panels as from 2020.

This was coupled with a scheme providing free panels in areas receiving most sunlight.

A flurry of legislation signed off in 2018 by the former governor, Jerry Brown, affected almost every part of the state's energy and transport sectors, with new incentives for the use of renewables and a target date of 2045 for the state to be carbon neutral.

The perhaps predictable response of the California Manufacturers & Technology Association was to warn that the measures could damage an economy that was by itself the fifth largest in the world.

Their spokesman, Gino DiCaro, said sweeping green reforms encouraged companies 'to make long-term manufacturing investment decisions in other places'.

With every new oil well sunk, with every mine opened, with every fresh fracking scheme rubber-stamped, with every new airport runway approved, the day of reckoning draws inexorably closer.

And yet, as we've seen, there are solutions to hand if we can persuade the makers and shakers to use them.

Mapping a rescue

In September 2019, 55 authors from across academia, industry, policy and consultancy, published what they called their Exponential Climate Action Roadmap in Stockholm.

Their report included 36 actions required to halve greenhouse gas emissions by 2030 to meet the 1.5°C Paris climate ambition.

It ran through many of the technical and social solutions outlined in these pages – from solar and wind energy to a reduced consumption of red meat – but also placed an onus on the man and woman in the street: strong civil society movements were needed, it stated, to ramp up the pace of change.

> 'We are the first generation to feel the sting of climate change, and we are the last generation that can do something about it.'
>
> *Jay Inslee, Governor of Washington*

'I see evidence that social and economic tipping points are aligning,' said Christiana Figueres, a former top climate official at the UN. 'We can now say the next decade has the potential to see the fastest economic transition in history.'

Meanwhile Manuel Pugal-Vidal of the World Wildlife Fund stressed the need for fairness in tackling the crisis.

'Governments must introduce national targets to reach net-zero emissions of 50 per cent by 2030,' he said. 'Immediate removal of fossil-fuel subsidies is a priority. Yet policies must be equitable and fair or risk failure.

'Developed nations with significant historic emissions also have a responsibility to reduce emissions faster. Cities and states – not only countries – will also be important change makers.'

There was a degree of optimism from Johan Rockström, director of the Potsdam Institute for Climate Impact Research in Germany.

'This is a race against time,' he said, 'but businesses and even entire industries have made many significant transitions in less than ten years.'

The food on our plates

At the Exponential Climate Change Roadmap launch, Brent Loken from the non-profit EAT Foundation warned that the food and agriculture sector was 'the dark horse' in the fight against climate change, and possibly the hardest in which to achieve a rapid halving of emissions.

- **There needed to be a switch from the high consumption of red meat and ultra-processed foods to a more healthy diet of fruit, legumes and vegetables.**

- **More sustainable farming practices globally were hampered by contradictory subsidies, poor land-use planning, insufficient funding and a focus on quick profits.**

Asking the people

When the Isle of Man devised a Climate Change Mitigation Strategy 2020-2030, it asked the people what they thought.

- There was overwhelming support for new buildings being constructed to a 'nearly zero emissions' standard; for community renewable energy; for onshore wind generation; and for the installation of renewables before 2030.
- 79 per cent (16 per cent against) were prepared to pay a climate change levy.
- 62 per cent (26 per cent against) supported a levy on new fossil fuel boilers.
- 47 per cent (40 per cent against) supported a supplement on road tax for fossil fuel vehicles.
- 49 per cent would pay more than 2 per cent of household income on energy efficiency (10 per cent would pay none).

Two decisions already made on the island: gas boilers will be banned in new homes by 2025 and no petrol and diesel cars will be sold after 2040.

THE DOOMSDAY DECADE

> 'We do not inherit the earth from our
> ancestors – we borrow it
> from our children.'
>
> *Native American proverb*

A Citizens' Assembly

A prime example of active public involvement in combating climate change as advocated by the Exponential Roadmap authors was the first ever Climate Emergency People's Assembly, held in the foyer of the Welsh parliament building in Cardiff in July 2019.

Initiated by Extinction Rebellion, and sponsored by two Welsh Assembly members, it set out to explore the role of direct civic engagement in developing 'fair and effective' priorities towards speedily cutting the principality's carbon emissions.

The hope was eventually to form a 'Citizens' Assembly', with participants from across society having a direct and formal role in decision-making, while an immediate proposal was that a climate emergency plan for Wales should include a clear call to 'get on with it'.

The Welsh Assembly initiative rapidly inspired councils large and small throughout the world to declare their own state of climate emergency.*

By the end of the year well over a thousand local governments in 23 countries – representing between them almost 300 million people – had made the declaration, with a commitment to drive down fossil fuel emissions at emergency speed.

Zero Carbon Britain

The lead speaker at the Cardiff event was Paul Allen from the Centre for Alternative Technology (CAT), an eco-centre set in an abandoned slate quarry in Powys, mid-Wales, whose visitor attractions include a water-balanced funicular; solar, hydropower and wind power; and a site-wide electricity grid powered by renewable energy.

---

* *An online map plots the take-up of the initiative, with 80 per cent of the British population already living in climate emergency areas by the end of 2019.*

CAT's Zero Carbon Britain Hub and Innovation Lab was established in 2019 in response to requests from governments, councils, community groups and businesses for help in drafting policies to turn their bald climate emergency declarations into realistic on-the-ground action plans tailored to their individual needs.

That realism, CAT acknowledges, means addressing a wide range of obstacles: 'We need solutions that work across a complex range of interacting areas – solutions that not only offer technical fixes but also help overcome political, cultural, economic and psychological barriers.'

We'll leave a last clarion call to Paul Allen as we negotiate the manifest perils of this Doomsday Decade:

'The technology says we can; the science says we must; and now it's time to say we will.'

Glossary

Algae Simple green plants living in water and wet places, from single cells to large seaweeds.

Anaerobic Existing in the absence of oxygen.

Anthropocene Relating to the current geological age during which human activity has been the dominant influence on climate and the environment – hence **anthropogenic** climate change.

Atmosphere The gases surrounding a planet or moon. The Earth's is a mixture of gases called air.

Biomass Living biological material, such as plants, used to make a solid, liquid or gas fuel.

Carbon capture and storage A method for trapping carbon dioxide, stopping it from being released into the atmosphere and locking it away permanently.

Carbon cycle The processes by which carbon compounds circulate in the environment, entering living tissue by photosynthesis and returning to the atmosphere through respiration, the decay of dead organisms and the burning of fossil fuels.

Carbon footprint The amount of carbon released into the atmosphere via the activities of an individual, an organisation or a community.

Carbon offsetting A scheme which balances one individual's or organisation's carbon dioxide emissions with a lower output from another source.

Carbon sink An ocean or forest capable of absorbing carbon dioxide from the atmosphere.

GLOSSARY

Carbon trading The buying and selling of carbon credits with the aim of reducing the amount of carbon dioxide released into the atmosphere.

Divestment The process of selling off investments. In regard to climate change, the principled selling of shares in fossil fuel companies.

Fossil The remains or imprint of a plant or animal that lived in prehistoric times, now embedded in rock.

Fossil fuel A substance that can be used as a source of energy, usually by burning it.

Glacier A slowly moving mass or river of ice.

Global warming An increase in the temperature of the Earth's atmosphere, a term usually implying human causes.

Greenhouse effect The process by which the atmosphere absorbs energy radiated from the Earth's surface as a result of solar heating.

Greenhouse gas A gas that produces the greenhouse effect. The main greenhouse gases in the Earth's atmosphere are methane, carbon dioxide, water vapour and nitrous oxide.

Gulf stream A warm ocean current that originates in the Gulf of Mexico and follows the eastern coasts of the US and Newfoundland before crossing the Atlantic Ocean.

Jet stream One of several fast winds blowing around the world 13–20km (8–12miles) above ground.

LED A light-emitting diode, used in low energy light bulbs.

Nuclear fission A nuclear reaction in which a large nucleus splits in two and releases energy.

Nuclear fusion A nuclear reaction in which two small atomic nuclei fuse together to form a bigger nucleus and release energy.

Ocean acidification The decrease in pH of the seas caused by the uptake of carbon dioxide from the atmosphere.

Permafrost Ground so cold that it is permanently frozen.

pH A figure expressing the acidity (low) or alkalinity (high) of a solution on a scale of which 7 is neutral.

Photosynthesis The process by which plants use sunlight to synthesise nutrients from carbon dioxide and water, giving off oxygen as a byproduct.

Renewable energy Energy obtained from natural sources, such as the Sun or wind, which is constantly produced and is not used up like coal or oil.

Solar energy Energy, usually heat or electricity, obtained from sunlight.

Tidal barrage A structure built to extract energy from tides for generating electricity.

Tokamak An apparatus for producing fusion reactions in hot plasma.

Turbine A machine with a part that spins when a liquid or gas hits its vanes or blades. Turbines extract energy from moving liquids and gases for powering machines such as generators for producing electricity.

Climate Change Timeline

1760 Scottish chemist Joseph Black discovers carbon dioxide.

1776 Alessandro Volta discovers methane.

1793 Joseph Priestley discovers nitrous oxide.

1814 An elephant walks on the ice of the Thames during London's last Frost Fair.

1816 'Year without a summer' after Mount Tambora volcano erupts.

1830 World's first intercity railway service, between Liverpool and Manchester, triggers intensification of the industrial revolution.

1840 Christian Friedrich Schönbein discovers ozone.

1853 First commercial oil well drilled, in Poland.

1859 First oil well drilled in the US. Irish physicist John Tyndall discovers that gases in the atmosphere can cause global warming.

1870 Level of CO_2 in the atmosphere is 290ppm.

1878 Cragside in Northumberland is the first house to have its lighting powered by hydroelectricity.

1896 Svante Arrhenius is first person to describe a man-made greenhouse effect caused by CO_2 produced by coal burning.

1900 Ferdinand Porsche builds the first petrol-electric hybrid car.

1911 First geothermal power station opens in Lardello, Italy.

1930s Milutin Milanković suggests that ice ages are caused by changes in the Earth's orbit.

1934 Italian physicist Enrico Fermi achieves the first artificial nuclear fission, leading to the development of nuclear power.

1942 Enrico Fermi and Leo Szilard create the first man-made nuclear reactor, named Chicago Pile-1.

1947 *The Bulletin of the Atomic Scientists* in Chicago launches its Doomsday Clock.

1951 The experimental nuclear reactor, EBR-1 in Arco, Idaho, is the first to generate electricity.

1954 The Obninsk Nuclear Power Plant in Russia is the first to supply electricity to the grid.

1956 Opening of the first commercial nuclear power station at Calder Hall in Sellafield, England. Beginning of nuclear fusion research using the Russian Tokamak reactor design.

1957 Windscale and Kyshtym nuclear accidents. Roger Revelle finds that carbon dioxide produced by human activities is not readily absorbed by the oceans.

1958 Charles Keeling measures CO_2 in the atmosphere and detects an annual rise in its levels.

1961 SL-1 nuclear accident.

1962 The biologist and conservationist Rachel Carson publishes her seminal book *Silent Spring*.

1965 A climate change conference is held in Boulder, Colorado.

1966 The Rance tidal power plant opens in France.

CLIMATE CHANGE TIMELINE

1968 Climate studies suggest the possibility of a collapse of the Antarctic ice sheets, which would cause a catastrophic rise in sea levels.

1970 The first Earth Day, designed to inspire a greater awareness of the environment.

1971 A SMIC (Study of Man's Impact on Climate) conference warns of the danger of rapid and serious global climate change caused by humans.

1973 An oil crisis encourages interest in alternative forms of energy.

1974 The chemists Sherwood Rowland and Mario Molina discover that the CFCs in aerosol sprays are destroying the Earth's protective ozone layer.

1975 Syukuro Manabe and others create computer models that predict doubling the amount of CO_2 in the atmosphere will raise temperatures by several degrees.

1979 Three Mile Island nuclear accident.

1986 Chernobyl nuclear accident.

1987 The Montreal Protocol agrees restrictions on the release of gases that harm the ozone layer.

1988 The Intergovernmental Panel on Climate Change (IPCC) is established. The British prime minister, Margaret Thatcher, is the first major world leader to call for action on climate change.

1989 The Global Climate Coalition is set up by a group of mainly US businesses to oppose actions to reduce greenhouse gas emissions.

1992 The Rio Earth Summit.

1995 The IPCC warns that serious global warming is likely in the next century.

1997 The Kyoto Protocol agrees targets for reducing greenhouse gas emissions. Toyota introduces the Prius, the first mass market hybrid car.

1999 The 1990s are declared the warmest decade in a thousand years.

2001 President George W. Bush refuses to sign up to the Kyoto Protocol because it will harm the US economy. The IPCC warns of 'unprecedented' global warming with possibly severe effects.

2003 Tens of thousands die during a summer heatwave in Europe.

2005 New Orleans is devastated by Hurricane Katrina. Cadarache, France, is chosen as the site for a new nuclear fusion research reactor, ITER.

2006 Al Gore's film about global warming, *An Inconvenient Truth*, is released.

2007 IPCC's fourth report warns that the world's ice sheets are shrinking faster than expected. IPCC and Al Gore jointly win the Nobel Peace Prize.

2008 55 billion tons of ice break away from the Wilkins Shelf on the Antarctic coast. The UK is the first country in the world to introduce a Climate Change Act.

2009 The Maldives government holds the world's first underwater cabinet meeting to highlight the rise in sea levels because of climate change.

2015 The Paris Agreement, signed by almost 200 nations, aims to restrict the increase in global temperatures

CLIMATE CHANGE TIMELINE

to 1.5°C (2.7F) above pre-industrial levels through reductions in greenhouse gas emissions. Pope Francis issues his encyclical *Laudato si'*, lamenting environmental degradation and global warming.

2016 A boy dies in an anthrax outbreak in Russia's Arctic Circle after thawing of the permafrost exposes ancient animal remains containing the deadly bacteria.

2017 Al Gore's second film is released, *An Inconvenient Sequel: Truth to Power*.

2018 The campaigning group Extinction Rebellion is founded. Greta Thunberg stages her first school strike, which is immediately copied internationally.

2019 'Funerals' of the Ok glacier in Iceland (where global CO_2 in the atmosphere is recorded as 415ppm) and the Pizol glacier in Switzerland. President Trump says the US will withdraw from the Paris climate accord. Brazil's president Bolsonaro takes office and rescinds his country's offer to host that year's climate change conference: Greta Thunberg attends the conference (in New York) and tells delegates 'You have stolen my dreams'. Hurricane Dorian devastates the Bahamas. The Royal Shakespeare Company ends its sponsorship deal with BP after criticism by young theatre-goers. The Centre for Climate Repair opens at Cambridge University in England. An Exponential Climate Change Roadmap is published in Stockholm, promoting policies that would halve greenhouse gas emissions by 2030. A People's Assembly in Wales inspires councils worldwide to declare a state of climate emergency.

Index

A
Abidjan 75
Adam, Shiham 49
Adani Carmichael 138–140
Africa 63, 160–162
Agriculture 57–58, 98, 171
Akasaki, Isamu 148
Algae 50, 79
Allen, Paul 174–175
Allison, Wade 153
Amano, Hiroshi 148
American Chemistry Council 105
An Inconvenient Truth 97
Anaerobic bacteria 20
Andes, The 43
Antarctic 47, 74, 133
Anthrax 43
Anthropocene, The 20
Araújo, Ernesto 123
Arctic 43, 44, 47, 74
Argentina 71, 95
Arrhenius, Svante 18
Assman, Richard 27
Atmosphere 26
Attenborough, Sir David 106, 109–110
Australia 50, 63, 71, 110–111, 138–141, 160

B
Bahamas 60
Bahrain 45
Baltic Sea 14
Bangkok 75
Bangladesh 143
Banks, Arron 96
BBC 107
Beddington, John 87
Belgium 143
Benedict, Pope 114
Bermuda 45
Berners-Lee, Mike 141
Bezos, Jeff 154
Bhutan 71
Bioenergy 77, 150–151
Biological Conservation 81
Bloomberg, Michael 154
Bolsonaro, Jair 55–57, 123
Boston, Massachusetts 75
BP 139
Brazil 53, 55–57, 63, 71, 73, 95, 123, 141, 149, 151
Brown, Jerry 168
Bulletin of the Atomic Scientists 72
Bush, George W 112
Byron, Lord 21

INDEX

C

California 168
Canada 42, 45, 71, 149
Canadell, Pep 73
Cap and trade 136
Carbon bubble theory 134
Carbon capture 156–158
Carbon cycle 31–32
Carbon dioxide 9, 11, 17–19, 31–32, 34, 53, 67, 76, 85, 89, 113, 120
Carbon footprint 56
Carbon offsetting 137
Carbon sinks 34
Carney, Mark 134
Carson, Rachel 98
Centre for Alternative Technology 174–175
Centre for Climate Repair 155
CFCs 132–133
Chennai 75
Chernobyl 150
Chile 71
China 13, 21, 60, 75, 141, 149, 154
Climate Accountability Institute 90
Climate Action Tracker 70
Climate Change Act 12, 123
Climate Change – the Facts (film) 110
Climate Emergency Declarations 173–174
Climate Focus 53
Coal 110–113, 135, 140, 147, 164
Committee on Climate Change (UK) 12, 129, 131, 142
Coral reefs 50–52
Costa Rica 71, 151
Cragside, Northumberland 150
Crowther, Tom 159
Cuadrilla 136
Culham Science Centre 153
Cyclones 60
Cyprus 47

D

Das Wetter 27
DDT 98
De Burt, Leon 27
Denmark 14
DiCaro, Gino 168
Diffenbaugh, Noah 85
Diseases 82–84
Divestment 138–140
Dogger Bank 146
Doomsday Clock 72
Doran, Gregory 139
Dorian (hurricane) 60
Drake, Edwin 22
Draper, Charles 46
Drax power station 147

Drought 60, 63, 87–88, 168
D'Souza, Dinesh 96

E
Earth Overshoot Day 45
Ecologist, The 161
Ecuador 62, 75
Edison, Thomas 148
El Niño 62–63
El Salvador 151
Electric cars 114, 142
Engel, Ditler 163
Environmental Protection Agency 104–105
Ethiopia 71
European Union 71, 86
Evans, Mel 115
Exosphere 27
Exponential Roadmap 169
Extinction Rebellion 80, 173
Eyjafjallajökull volcano 41

F
Farming 57–58
Figueres, Christiana 170
Flatulence in cattle 57–58
Floods 44, 46–49, 60, 74–76, 87
Florida 76
Foer, Jonathan Safran 127

Fossil fuels 8, 18, 51, 61, 90, 108, 134, 164–165
Fox News 96, 103
Fracking 136
France 12, 95
Francis, Pope 113–116
Friedman, Thomas 99–100
Frost fairs 14
Fukushima 150, 152

G
Gambia, The 71
Gansu wind farm 146
Gates, Bill 154
Geothermal energy 151
Germany 41, 95
Glacier National Park 42
Glaciers 39–43
Global Environmental Change 51
Global warming 15, 18, 34, 61, 101
Goldsmith, Teddy 161
Goodall, Jane 117
Gore, Al 4, 97–99
Graveney, Kent 144
Great Barrier Reef 50, 110
Great Dying, The 19
Great Green Wall 160
Great Ocean Conveyor 36–37
Great Pacific Garbage Patch 52

INDEX

Green New Deal 99–106
Green taxes 137
Greenhouse effect 28–29
Greenhouse gases 25, 30–31, 66–67, 70, 73, 85, 90, 98, 169
Greening the oceans 155
Greenland 47, 74
Greenpeace 136
Griffith, Saul 128
Grolar bears 42
Guangzhou 75
Guayaquil 75
Guardian style guide 101
Gulf Stream 36
Guterres, António 72

H

Haigh, Joanna 65
Hammond, Philip 129–131
Hansen, James 76
Harry, Duke of Sussex 116–117
Hawaii 9
Heatwaves 60, 84
Heede, Richard 90
Hill, Tessa 52
Himalayas 42
Ho Chi Minh City 75
Hodel, Donald 133
Howe, Cymene 40
Hurricanes 60
Huskisson, William 7
Hydroelectricity 149–150

I

Ice ages 13–15
Icebergs 44
Iceland 39, 41, 151
Idai (cyclone) 60
India 13, 21, 63, 71, 75
Indiana University 1165
Indonesia 47, 54, 62, 71, 75
Industrial revolution 7–11
Ingraham, Laura 96
Insects 79, 81–82
International Space Station 26
IPCC 10, 70, 97, 151
Ireland 21
Islamic State (ISIS) 88
Isle of Man 172
Italy 41
Iter 154
Ivory Coast 75

J

Jakarta 47, 75
Japan 71, 73, 75 143
Jenatzy, Camille 142

K

Katrina (hurricane) 103
Kazakhstan 71
Kebnekaise 41
Keeling, Charles 9
'Keep it in the Ground' 134

Kelley, Colin 88
Kenya 151
Kigali Amendment 133
King, Sir David 65, 158
Kiribati 48
Klein, Naomi 101–106
Knowles, Michael 96
Kolkata 75
Kuwait 45
Kyoto Agreement 98

L

La Niña 63
Lagarde, Christine 125
Lake Maggiore 31
Laudato si 115
Lavoisier, Antoine 33
Lawson, Lord 123
Leadsom, Andrea 147
LED light bulbs 148
Lekima (typhoon) 60
Little Ice Age 14
Loken, Brent 171
London Array 145
Łukasiewic, Ignacy 22
Luntz, Frank 112
Luxembourg 45

M

Maamau, Taneti 48
Madagascar 60
Maeslant barrage 59
Magnason, Andri Snær 40
Malawi 60

Maldives, The 48–49
Manhattan Harbour 14
Mann, Michael 91
Mass extinctions 19
Mauna Loa 9, 101
McKibben, Bill 134–135
Mesosphere 26
Methane 3, 44, 57–58
Mexico 71
Miami 75
Milanković, Milutin 15–16
MIT 154
Molina, Mario 132
Monbiot, George 106–108
Mongolia 45
Montreal Protocol 133
Moon, The 25
Morocco 47
Morrison, Scott 110
Mosquitoes 82–83
Mozambique 60
Mumbai 75
Mustill, Tom 108

N

Nagoya 75
Nakamura, Shuji 148
Nasheed, Mohammed 48
Nature Climate Change 75
Nature Geosciences 67
Netherlands, The 12–14, 143
New Orleans 75

INDEX

New York 75
New Zealand 71
Nobel prizes 97, 133, 148
Noor Abu Dhabi solar plant 144
Norilsk 44
North Atlantic Drift 36
Norway 12, 71
Nuclear fission 150–152
Nuclear fusion 153–154

O

Ocasio-Cortez, Alexandria 100
Ocean acidification 51–52
Oil 22, 90–91, 113–114, 135, 139, 164
Ok Glacier 39–40
On Fire: the Burning Case for a Green New Deal 103
Oxygen 32, 109
Ozone layer 132–133

P

Pakistan 47
Paris Agreement 11
Pastoruri Glacier 43
Pence, Mike 122
Permafrost 43–44
Peru 43, 62, 71
Philippines, The 71, 151
Photosynthesis 32
Pizol Glacier 41
Plastics 52
Poland 22
Polo, Marco 22
Priebus, Reince 122
Priestley, Joseph 33
Proceedings of the National Academy of Sciences 88
Puffins 79
Pugal-Vidal, Manuel 170
Putin, Vladimir 122

Q

Qatar 45

R

Rain forests 53–57, 77, 108
Reagan, Ronald 133
Red List (species) 78
Reforestation 158–162
Refreezing the poles 155
Renewables 141–154
Rhone Glacier 41
Rice growing 57
Rice University 40
Roberts, Callum 156
Roberts, Debra 70
Rockström, Johan 171
Romm, Joseph 44
Rotterdam 59
Rowland, Sherwood 132
Royal Shakespeare Company 139
Russia 43–44, 71, 122–123, 141, 149
Rwanda 143

S

Sahara Desert 160
Sand eels 79
Sankt Gallen canton 41
Saudi Arabia 71, 73
Science Advances 58
Sellafield 152
Shenzhen 75
Siberia 43
Siferis, Lisa 116
Silent Spring 98
Singapore 71
Skirball fire, LA 112
Smith, 'Uncle Billy' 22
Soda water 33
Solar energy 114, 141–145
South Africa 71, 73
South Korea 71, 143
Sri Lanka 63
Stanford University 85
Stephenson, Robert 7
Storm surges 59
Stratosphere 26–27
Streck, Charlotte 54
Styring, Peter 156
Surat 75
Swan, Joseph 148
Swansea Bay 149
Sweden 14, 41
Switzerland 14, 41, 71
Syria 88

T

Taft, William 42
Tampa, Florida 75
Tata Steel 156
Tengger solar park 144
Thailand 75
Thames Barrier 46
There Is No Plan B 141
Thermosphere 26
Three Mile Island 150
Thunberg, Greta 4, 93–96, 108, 111
Tianjin 75
Tokamak Energy 154
Tomales Bay 52
Tong, Anote 48
Trade winds 62
Trinidad and Tobago 45
Troposphere 26, 35
Trump, Donald 4, 11, 73, 96, 104–105, 122
Tsunamis 59
Tundra 44
Tunisia 47
Turkey 71, 95
Typhoons 60

U

UAE 45, 71
Ukraine 71
United Nations 11, 72
University of California 52

INDEX

USA 11, 21–22, 42, 45, 52, 71, 73, 75, 104–105, 141, 149, 151, 154, 160

V
Vidal, John 117
Vietnam 47, 75
Viner, Katharine 101
Vogue magazine 117
Volcanoes 17, 20, 31, 41
Volta, Alessandro 31

W
Walney Extension 145
Wanless, Harold 76
Watson, Robert 81
Wave and tidal energy 146–149
'We Are Still In' 11
Weather 17, 27
Wellington, Lord 7
West Atlantic Ice Sheet 74
Western Arctic Ice Sheet 47, 74
Wildfires 60, 112, 168
Wildlife 77–81
Wilkins Shelf 47
Wind power 145–146
Windscale 150, 152
World Health Organisation 84
World Meteorological Organization 73
World Wildlife Fund 51, 78

X
Xiamen 75

Y
Yameen, Abdulla 49
Year Without a Summer, The 21

Z
Zero Carbon Britain 174–175
Zhanjiang 75
Zimbabwe 60

Cherished  Library

Some other
Very Peculiar Histories®

The Blitz
David Arscott
ISBN: 978-1-907184-18-5

Castles
Jacqueline Morley
ISBN: 978-1-907184-48-2

Charles Dickens
Fiona Macdonald
ISBN: 978-1-908177-15-5

Golf
David Arscott
ISBN: 978-1-907184-75-8

Great Britons
Ian Graham
ISBN: 978-1-907184-59-8

Ireland
Jim Pipe
ISBN: 978-1-905638-98-7

Kings & Queens
Antony Mason
ISBN: 978-1-906714-77-2

London
Jim Pipe
ISBN: 978-1-907184-26-0

Make Do and Mend
Jacqueline Morley
ISBN: 978-1-910184-45-5

Scotland
Fiona Macdonald

Vol. 1: From ancient times to Robert the Bruce
ISBN: 978-1-906370-91-6

Vol. 2: From the Stewarts to modern Scotland
ISBN: 978-1-906714-79-6

Wales
Rupert Matthews
ISBN: 978-1-907184-19-2

Whisky
Fiona Macdonald
ISBN: 978-1-907184-76-5

Visit
www.salariya.com
for our online catalogue and
free fun stuff.